Osteocardiology

Nalini M. Rajamannan

Osteocardiology

Cardiac Bone Formation

Nalini M. Rajamannan
Most Sacred Heart of Jesus Cardiology and Valvular Institute
Sheboygan, Wisconsin
USA

ISBN 978-3-319-64993-1 ISBN 978-3-319-64994-8 (eBook)
DOI 10.1007/978-3-319-64994-8

Library of Congress Control Number: 2017955376

Printed on acid-free paper

This Springer imprint is published by Springer Nature
The registered company is Springer International Publishing AG
The registered company address is: Gewerbestrasse 11, 6330 Cham, Switzerland

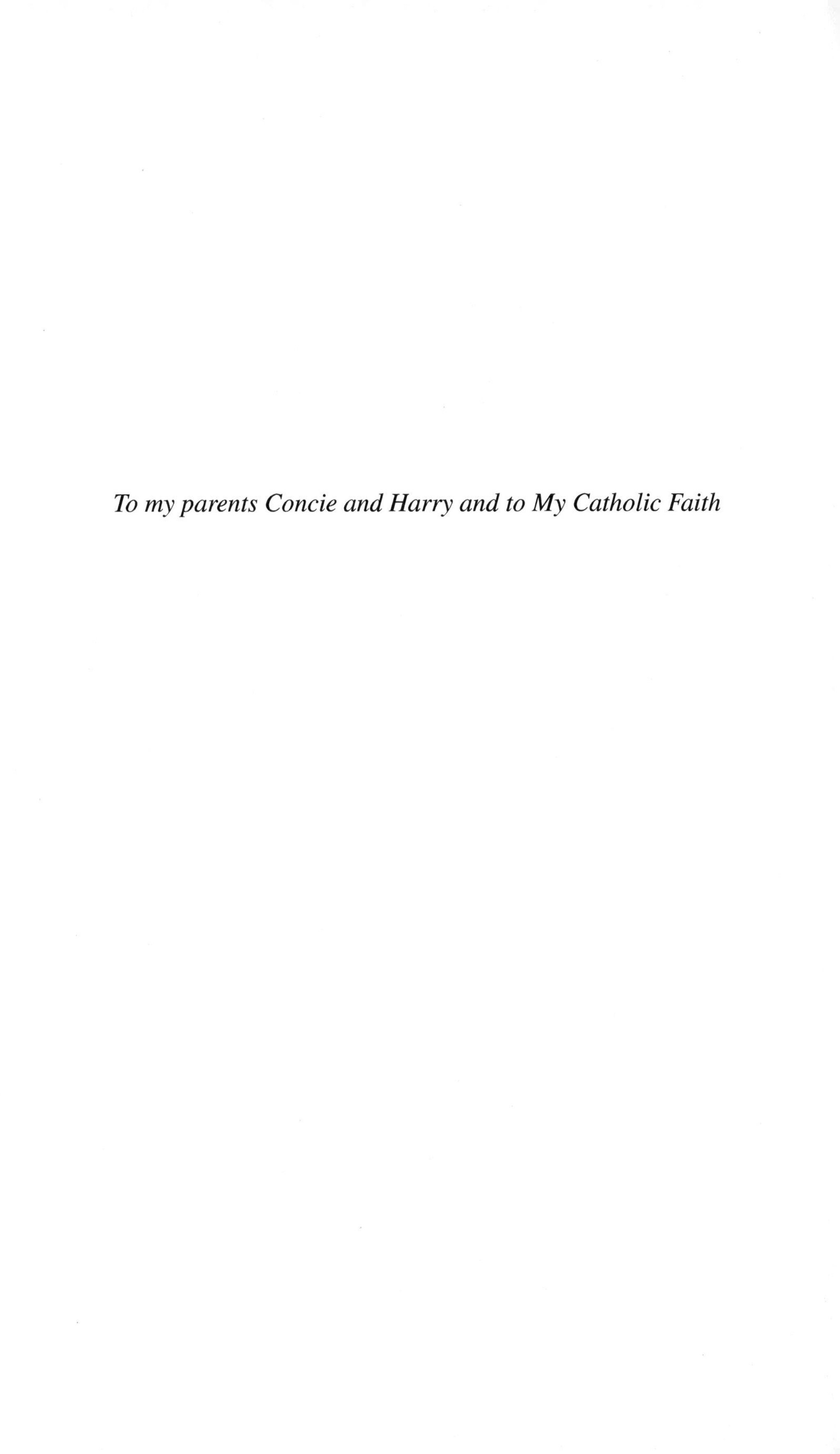

To my parents Concie and Harry and to My Catholic Faith

Foreword

In 1977, the National Heart, Lung, and Blood Institute (NHLBI) issued the first of several clinical practice guidelines and position papers, which has provided global leadership for a research, training, and education program to promote the prevention and treatment of heart, lung, and blood diseases.

Guidelines from the National High Blood Pressure Education Program, the National Cholesterol Education Program, the Calcific Aortic Valve Disease, etc., includes accumulation of the research and experts in the fields, cholesterol, blood pressure, coronary artery disease and calcific aortic valve disease. Over the years, health care systems and providers have used these guidelines and position papers for the prevention, detection, evaluation, and treatment of cardiovascular disease risk factors, and lung and blood diseases.

NHLBI convened expert panels to update the existing clinical guidelines on cholesterol, blood pressure, and overweight/obesity, by conducting rigorous systematic evidence reviews. At the same time a working group—in the field of calcific aortic valve disease—helped to define the concept of an active biologic process in the aortic valve. The driving force for these guidelines and working groups was the recognition that despite the enormous progress over the last 60 years, cardiovascular disease remains the leading cause of death in the United States.

Since implementing the new collaborative partnership model for developing guidelines based upon NHLBI-sponsored systematic evidence reviews, working groups have worked successfully with the American Heart Association (AHA), the American College of Cardiology (ACC), and other professional societies to develop new cardiovascular disease prevention. The new guidelines—published in 2017 by the AHA, and ACC, and endorsed by other professional societies—provide a valuable updated roadmap to help clinicians and patients manage CVD prevention and treatment challenges.

Osteocardiology has evolved from the ground-breaking research in the field of basic science, observational studies, clinical trials, large cohort databases, and finally, the work of the NHLBI to bring consensus panels together as leaders in the field to fight and win the battle of cardiovascular calcification, globally.

Wisconsin, USA Nalini M. Rajamannan, M.D.

Preface

Osteocardiology is an exciting and new field of science, which will become the cornerstone for defining the timing and treatment of cardiovascular calcification in the future. In 2017, with the advent of large cohort databases, and experimental mechanistic studies, research has elucidated evidence confirming that traditional cardiovascular risk factors are responsible for the development of atherosclerotic calcification. Over the past 50 years, experimental studies have identified the critical elements of atherosclerosis, including foam cell formation, vascular smooth muscle cell proliferation and extracellular matrix synthesis, which over time forms bone in the heart. The Nobel Prize in Physiology or Medicine 1998 was awarded jointly to Drs. Robert F. Furchgott, Louis J. Ignarro and Ferid Murad, "for their discoveries concerning nitric oxide as a signaling molecule in the cardiovascular system". This discovery helped to understand the role of the endothelium in normal vasodilatation for maintenance of normal vascular biology, and the effect of oxidative stress on the down-regulation of endothelial nitric oxide synthase in the development of disease.

In the mid 1990s Dr. Linda Demer pioneered the concept of bone biology in the vascular calcification and developed sophisticated models of oxidative stress and bone analysis. Dr. Catherine Otto and Dr. Kevin O'Brien, from University of Washington, began early studies in identifying the presence of lipoproteins in the calcifying valves. Dr. Emile Mohler, at the same time, identified bone formation in the calcifying aortic valve. In vitro, Dr. Christopher Johnson developed cell culture models of the aortic valve, performing pioneering work in the field of fibronectin synthesis. Dr. Thomas Spelsberg's collaboration and friendship, was instrumental in the merging of bone biology and valve biology. Dr. Philippe Sucosky, in bioengineering, has become the leader in the field of valve hemodynamics, in addition to the seminal contributions from Dr. Simmons in endothelial gene expression, Dr. Bob Weiss studying aging mouse models of CAVD, Dr. Pibarot in echo hemodynamics of valvular heart disease, Dr. Aikawa, the first to demonstrate a renal failure model of aortic valve disease with bone loss, Dr. Genest studies in patients with Familial Hypercholesterolemia has defined cardiovascular calcification phenotype, and Dr. Berger-Klein, defining the role of biomarkers in CAVD. Finally, the bench to beside seminal study, RAAVE performed by Dr. Luis Moura from Porto, Portugal will set the stage for future clinical trials in the field of cardiovascular calcification.

MESA, Multi Ethnic subclinical atherosclerosis cohort, has developed a powerful database of CT imaging of a population which does not have clinical overt cardiovascular disease, making this tool of great importance in advancing our knowledge of the timing of calcification and the risk factors associated with this phenotype. Many of the several discoveries from this database will be outlined in this textbook of bone formation in the heart.

My 30 years of experience in the field of valve biology, echocardiography, clinical trials, and a MESA researcher, this overview of bone formation in the heart will hopefully, become the cornerstone to educate medical students, residents, fellows, graduate students, physician scientists and scientists, for future research and ongoing development in medical therapies to slow or halt the progression of bone formation in the heart, i.e. osteocardiology.

Wisconsin, USA

Nalini M. Rajamannan, M.D.

Contents

Osteocardiology Risk Factors

1

Introduction

Osteocardiology, bone formation in the heart, has become a dominant field of scientific study due to the increasing sensitivity of imaging of the heart. Even though the incidence of coronary artery disease, is on the decline [1], it is still the number one cause of morbidity and mortality globally as reported by the World Health Organization [2]. The American College of Cardiology, the European Society of Cardiology, American Heart Association and the World Health Organization are leading the fight against heart disease, by improving the effort to diagnose and recognize subclinical disease before it causes major adverse cardiovascular events, such as stroke, heart attack, heart failure, and death.

Imaging modalities such as computed tomography of the chest and abdomen (CT), angiography, echocardiography, nuclear imaging and PET imaging have increased our knowledge in the diagnosis and detection of atherosclerosis, symptomatic and asymptomatic disease, and coronary artery calcification (CAC). Furthermore, these imaging modalities have increased our detection of calcific aortic valve disease (CAVD), mitral annular calcification, and thoracic and abdominal aortic calcification. Recent studies over the past 20 years have elucidated the epidemiology, anatomic localization and molecular biology signaling pathways critical in the development of osteocardiology—bone formation in the heart.

Atherosclerosis, is a systemic disease process in which fatty deposits, inflammatory cells and calcification build within the walls of arteries, valves and the endocardium of the heart specifically, the mitral annulus, as shown in Fig. 1.1. Atherosclerosis is responsible for the majority of cardiovascular events. Atherosclerosis can develop in a variety of end-organs, including the heart, brain, kidneys, and extremities. This chapter will outline the epidemiologic risk factors for the development of cardiovascular calcification and bone disease, to provide a foundation for this textbook entitled Osteocardiology.

N.M. Rajamannan, *Osteocardiology*, DOI 10.1007/978-3-319-64994-8_1

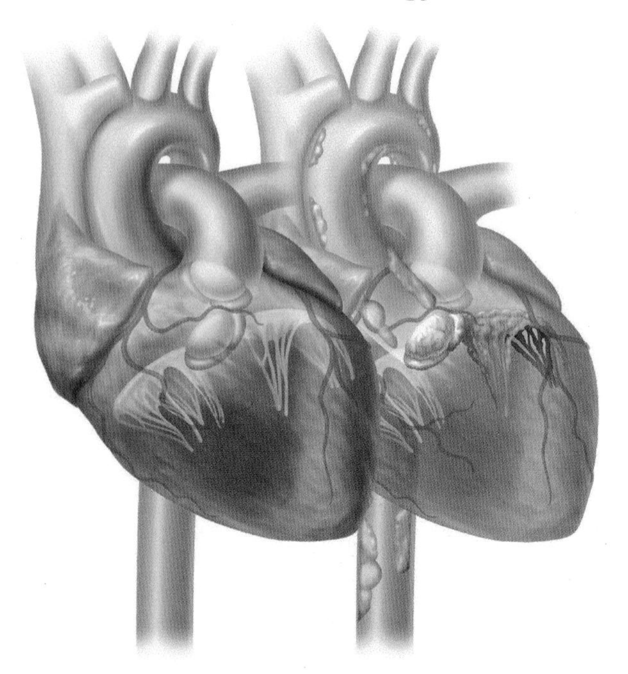

Fig. 1.1 Osteocardiology of the heart, coronary artery, aortic valve, mitral annular calcification and aortic calcification, from normal to diseased heart after long term exposure to osteocardiology risk factors

Framingham Risk Factor Database

Since the 1970s, the National Heart, Lung, and Blood Institute (NHLBI) have developed epidemiologic databases, which provide a foundation for the application of health science research. These cohort databases have provided critical understanding into the risk factors, outcomes and timing of medical therapy for a number of disease processes in the heart. In turn, the American Heart Association (AHA), American college of Cardiology (ACC) and NHLBI have written clinical guidelines, consensus documents, working group position papers, for the prevention, detection, and treatment of cardiovascular, lung, and blood diseases, over the last half a century. These working groups provide input into the development process for the next generation of clinical practice guidelines. For example, the Framingham 10-year risk score (Framingham Risk Score, or FRS) for coronary heart disease (CHD) risk assessment has elucidated the long term risk assessment for the diagnosis, evaluation, and treatment of high blood cholesterol [3]. The Framingham risk score (FRS) is a gender-specific algorithm used to estimate the 10-year cardiovascular risk of an individual. The FRS was first developed based on data obtained from the FHS, to estimate the 10-year risk of developing coronary heart disease [4]. The FRS is one of many to predict 10–30 year risk of developing cardiovascular disease. The scoring system is helpful in predicting long term risk of developing cardiovascular disease, but it also indicates who will likely benefit from preventive therapy, such as lowering blood pressure and lowering cholesterol etc. Over the past several decades, the AHA/ACC has developed 39 clinical guidelines, 66 expert consensus documents, and 18 performance measures, in the battle to diagnose and treat cardiovascular disease before it is too late.

Multi-Ethnic Study Atherosclerosis (MESA)

The Multi-Ethnic Study of Atherosclerosis was initiated in July 2000 to investigate the prevalence, correlates, and progression of subclinical cardiovascular disease (CVD) in a population-based sample of 6500 men and women aged 45–84 years. Six US Academic centers enrolled patients for the cohort database. The goal in 2000 was to recruit approximately 38% of the cohort are White, 28% African-American, 23% Hispanic, and 11% Asian (of Chinese descent). Baseline measurements included measurement of coronary calcium using computed tomography; measurement of ventricular mass and function using cardiac magnetic resonance imaging; measurement of flow-mediated brachial artery endothelial vasodilation, carotid intimal-medial wall thickness, and distensibility of the carotid arteries using ultrasonography; measurement of peripheral vascular disease using ankle and brachial blood pressures; electrocardiography; and assessments of microalbuminuria, standard CVD risk factors, socio demographic factors, life habits, and psychosocial factors. Blood samples will be assayed for putative biochemical risk factors and stored for use in nested case-control studies. DNA was extracted and lymphocytes will be immortalized for genetic studies. Measurement of selected subclinical disease indicators and risk factors were repeated for the study of progression over 7 years. Participants were followed through 2008, for identification and characterization of CVD events, including acute myocardial infarction and other coronary heart disease, stroke, peripheral vascular disease, and congestive heart failure; therapeutic interventions for CVD; and mortality [5].

The Multi-Ethnic Study of Atherosclerosis (MESA) has published over 1200 publications as of 2017, including characteristics of subclinical cardiovascular disease (disease detected non-invasively before it has produced clinical signs and symptoms) and the risk factors that predict progression to clinically overt cardiovascular disease or progression of the subclinical disease. The final MESA population-based sample is 6814 asymptomatic men and women aged 45–84. The final proportions of ethnic backgrounds in the cohort population, is approximately 38% white, 28% African-American, 22% Hispanic, and 12% Asian, predominantly of Chinese descent. MESA database has elucidated a number of clinical risk factors, which are important in the development of coronary artery calcification (CAC). These risk factors include increasing age, gender dependent, higher body mass index, higher blood pressure, abnormal lipids (higher low density lipoprotein or triglycerides, lower high density lipoprotein, or use of lipid-lowering medication), glucose disorders (impaired fasting glucose, untreated or treated diabetes mellitus), a familial history of CAC, chronic kidney disease (CKD), higher fibrinogen level and higher C-reactive protein level are more susceptible to CAC [6].

Cardiovascular Health Study

The Cardiovascular Health Study (CHS) is an NHLBI-funded observational study of risk factors for cardiovascular disease in adults 65 years or older. Starting in 1989, and continuing through 1999, participants underwent annual extensive clinical examinations. Measurements included traditional risk factors such as blood pressure and lipids as well as measures of subclinical disease, including echocardiography of the heart, carotid ultrasound, and cranial magnetic-resonance imaging (MRI). At 6-month intervals between clinic visits, and once clinic visits ended, participants were contacted by phone to ascertain hospitalizations and health status. The main outcomes are coronary heart disease (CHD), angina, heart failure (HF), stroke, transient ischemic attack (TIA), claudication, and mortality. Participants continue to be followed for these events. To date, more than 1300 research papers from CHS have been published and more than 300 ancillary studies are ongoing or complete.

The Cardiovascular Health Study (CHS) was critical for defining for the first time, using echocardiography to screen for aortic valve disease that, among adults >65 years, echocardiographically-detected aortic valve sclerosis was associated with a 50% increased risk of cardiovascular mortality [7]. In that study, aortic sclerosis also was associated with a 42% increase in risk of MI [7]. CHS investigators identified the clinical risk factors important for the development of atherosclerosis are also the independent risk factors for aortic valve stenosis including age, male gender, height (inverse relationship), history of hypertension, smoking and elevated serum levels of lipoprotein(a) and LDL levels [8].

Osteoporosis-Cardiovascular Risk Factors

Cardiovascular disease and osteoporosis are important causes of morbidity and mortality in the aging population. Osteoporosis or thinning of the bone is a complex heterogeneous disease, which is commonly associated with an increased incidence of atherosclerotic cardiovascular disease. For years these two disorders were thought to be unrelated, however, in the past several decades there is a growing number experimental [9] and epidemiological studies [10], which demonstrate a common pathophysiological and genetic risk factors [11, 12].

The possible link between cardiovascular disease and osteoporosis has been termed: "the bone-heart paradox." Studies in the field of atherosclerosis and osteoporosis have focused on the risk factor of oxidative stress as a mechanism for decreased bone formation in the bone and increased bone formation in the heart [13–15]. Vascular calcification and osteoporosis are both active biologic processes, which share common mechanisms including the BMP pathway, Wnt Pathway and OPG [16, 17].

The role of lipids in the vasculature is a well-known risk factor of atherosclerosis and bone formation in the heart [3, 40]. In the past two decades, traditional atherosclerotic risk factors have become known risk factors to activate the cellular mechanisms responsible for the bone-formation process in the heart. Investigators have described have described three types of calcification in the aorta: (1) atherosclerosis associated intimal calcification of the intima, (2) medial calcification/ Monckeberg type of sclerosis, and (3) genetic disorder–related calcification [3].

Framingham risk factors have demonstrated the role of lipids, hypertension, gender, body mass, renal disease and various lipoproteins, which are important in atherosclerotic heart disease [18]. Recently, similar risk factors for coronary artery disease and calcific aortic valve disease have recently been described including male gender, hypertension, elevated levels of LDL, and smoking [8, 19] which mimic those that promote the development of vascular atherosclerosis. Surgical pathological studies have demonstrated the presence of LDL and atherosclerosis in calcified valves, demonstrating similarities between the genesis of valvular and vascular disease and suggesting a common cellular mechanism [20, 21]. Patients who have the diagnosis of familial hypercholesterolemia develop aggressive peripheral vascular disease, coronary artery disease, as well as aortic valve lesions, which calcify with age [22–24]. Studies have also shown that the development of atherosclerosis occurs in the aortic valve in a patient with Familial Hypercholesterolemia with the low density lipoprotein receptor mutation [23]. The atherosclerosis develops along the aortic surface of the aortic valve and in the lumen of the left circumflex artery [23]. Calcification in the aorta has also been described in patients with Familial Hypercholesterolemia [24]. The work in the field of Familial Hypercholesterolemia provides a unique model of human disease, to study cholesterol metabolism, accelerated atherosclerosis and calcification in the heart.

American Heart Association Annual Statistics

The American Heart Association 2017 Heart statistics defines subclinical atherosclerosis in terms of biology, clinical diagnosis and long terms outcomes [1]. Numerous epidemiologic studies identified risk factors for calcific heart disease are similar to those of vascular atherosclerosis, including smoking, male gender, body mass index, hypertension, elevated lipid and inflammatory markers, metabolic syndrome and renal failure [7, 8, 19, 25–38]. For years this disease process was thought to be due to a degenerative phenomenon by which calcium attaches to the surface of the aortic valve leaflet and the lining of the vasculature. Understanding calcification, as the critical end-stage process which causes progression to severe stenosis and severe vascular occlusions leads to poor outcomes [39], is becoming important in the results of the randomized trials for treating cardiac calcification with medical therapy. In addition, there are a growing number of retrospective and prospective studies testing the hypothesis that atherosclerotic calcific AS may be targeted with medical therapy, however the randomized clinical trials are negative to date. This textbook will discuss the experimental evidence defining the cellular mechanisms of calcification in the heart and the translational implications of the current and future clinical trials testing medical therapies in the development of calcific heart disease.

In recent decades, advances in imaging technology have allowed for improved ability to detect quantity of atherosclerosis at all stages and in multiple vascular beds. Early identification of subclinical atherosclerosis could lead to more aggressive lifestyle modifications and medical treatment to prevent clinical manifestations of atherosclerosis such as myocardial infarction, stroke, or renal failure. Two modalities CT of the chest for evaluation of coronary artery calcification and B-model ultrasound of the neck for evaluation of carotid artery IMT, have been used in large studies with outcomes data and can help define the burden of atherosclerosis in individuals before they develop clinical events such as heart attack or stroke. Data on cardiovascular outcomes are beginning to emerge for additional modalities that measure anatomic and functional measures of subclinical disease, including brachial artery reactivity testing, aortic and carotid MRI, and tonometric methods of measuring vascular compliance or microvascular reactivity. This textbook will focus on the use of CT of the chest to measure calcification—the end-stage process of atherosclerosis in the cardiovascular system.

Summary

For decades, diagnosing calcification in the cardiovascular system has been elusive. The advent of computed tomography has opened the window to diagnosing calcification, and calculating the amount of calcification using the Agatston Score [38, 40, 41]. Understanding the hemodynamic and molecular mechanisms of calcification is critical towards understanding the end-stage calcified phenotype of atherosclerosis, and the specific anatomic locations of calcification in the heart. Osteocardiology provides the foundation for defining the timing and phenotype expression of bone formation in the heart. The osteocardiology theory correlates experimental evidence with hemodynamic calculations to define the cellular mechanisms of calcification to turn basic science into future clinical success.

References

1. Benjamin EJ, Blaha MJ, Chiuve SE, et al. Heart disease and stroke statistics-2017 update: a report from the American Heart Association. Circulation. 2017;135:e146–603.
2. WHO. The top 10 causes of death globally 2015. http://wwwwhoint/mediacentre/factsheets/fs310/en/2015.
3. National Cholesterol Education Program (NCEP) Expert Panel on Detection, Evaluation, and Treatment of High Blood Cholesterol in Adults (Adult Treatment Panel III). Third report of the National Cholesterol Education Program (NCEP) expert panel on detection, evaluation, and treatment of high blood cholesterol in adults (Adult Treatment Panel III) final report. Circulation. 2002;106:3143–421.
4. Wilson PW, D'Agostino RB, Levy D, Belanger AM, Silbershatz H, Kannel WB. Prediction of coronary heart disease using risk factor categories. Circulation. 1998;97:1837–47.
5. Bild DE, Bluemke DA, Burke GL, et al. Multi-ethnic study of atherosclerosis: objectives and design. Am J Epidemiol. 2002;156:871–81.

6. Kronmal RA, McClelland RL, Detrano R, et al. Risk factors for the progression of coronary artery calcification in asymptomatic subjects: results from the multi-ethnic study of atherosclerosis (MESA). Circulation. 2007;115:2722–30.
7. Otto CM, Lind BK, Kitzman DW, Gersh BJ, Siscovick DS. Association of aortic-valve sclerosis with cardiovascular mortality and morbidity in the elderly. N Engl J Med. 1999;341:142–7.
8. Stewart BF, Siscovick D, Lind BK, et al. Clinical factors associated with calcific aortic valve disease. Cardiovascular Health Study. J Am Coll Cardiol. 1997;29:630–4.
9. Rajamannan NM. Atorvastatin attenuates bone loss and aortic valve atheroma in LDLR mice. Cardiology. 2015;132:11–5.
10. Figueiredo CP, Rajamannan NM, Lopes JB, et al. Serum phosphate and hip bone mineral density as additional factors for high vascular calcification scores in a community-dwelling: the Sao Paulo Ageing & Health Study (SPAH). Bone. 2013;52:354–9.
11. Anagnostis P, Karagiannis A, Kakafika AI, Tziomalos K, Athyros VG, Mikhailidis DP. Atherosclerosis and osteoporosis: age-dependent degenerative processes or related entities? Osteoporos Int. 2009;20:197–207.
12. Freedman BI, Bowden DW, Ziegler JT, et al. Bone morphogenetic protein 7 (BMP7) gene polymorphisms are associated with inverse relationships between vascular calcification and BMD: the Diabetes Heart Study. J Bone Miner Res. 2009;24:1719–27.
13. Mody N, Parhami F, Sarafian TA, Demer LL. Oxidative stress modulates osteoblastic differentiation of vascular and bone cells. Free Radic Biol Med. 2001;31:509–19.
14. Parhami F, Tintut Y, Beamer WG, Gharavi N, Goodman W, Demer LL. Atherogenic high-fat diet reduces bone mineralization in mice. J Bone Miner Res. 2001;16:182–8.
15. Parhami F, Morrow AD, Balucan J, et al. Lipid oxidation products have opposite effects on calcifying vascular cell and bone cell differentiation. A possible explanation for the paradox of arterial calcification in osteoporotic patients. Arterioscler Thromb Vasc Biol. 1997;17:680–7.
16. Osako MK, Nakagami H, Koibuchi N, et al. Estrogen inhibits vascular calcification via vascular RANKL system: common mechanism of osteoporosis and vascular calcification. Circ Res. 2010;107:466–75.
17. Hjortnaes J, Butcher J, Figueiredo JL, et al. Arterial and aortic valve calcification inversely correlates with osteoporotic bone remodelling: a role for inflammation. Eur Heart J. 2010;31:1975–84.
18. Berry JD, Lloyd-Jones DM, Garside DB, Greenland P. Framingham risk score and prediction of coronary heart disease death in young men. Am Heart J. 2007;154:80–6.
19. Aronow WS, Ahn C, Kronzon I, Goldman ME. Association of coronary risk factors and use of statins with progression of mild valvular aortic stenosis in older persons. Am J Cardiol. 2001;88:693–5.
20. O'Brien KD, Reichenbach DD, Marcovina SM, Kuusisto J, Alpers CE, Otto CM. Apolipoproteins B, (a), and E accumulate in the morphologically early lesion of 'degenerative' valvular aortic stenosis. Arterioscler Thromb Vasc Biol. 1996;16:523–32.
21. Olsson M, Thyberg J, Nilsson J. Presence of oxidized low density lipoprotein in nonrheumatic stenotic aortic valves. Arterioscler Thromb Vasc Biol. 1999;19:1218–22.
22. Sprecher DL, Schaefer EJ, Kent KM, et al. Cardiovascular features of homozygous familial hypercholesterolemia: analysis of 16 patients. Am J Cardiol. 1984;54:20–30.
23. Rajamannan NM, Edwards WD, Spelsberg TC. Hypercholesterolemic aortic-valve disease. N Engl J Med. 2003;349:717–8.
24. Alrasadi K, Alwaili K, Awan Z, Valenti D, Couture P, Genest J. Aortic calcifications in familial hypercholesterolemia: potential role of the low-density lipoprotein receptor gene. Am Heart J. 2009;157:170–6.
25. Deutscher S, Rockette HE, Krishnaswami V. Diabetes and hypercholesterolemia among patients with calcific aortic stenosis. J Chronic Dis. 1984;37:407–15.
26. Hoagland PM, Cook EF, Flatley M, Walker C, Goldman L. Case-control analysis of risk factors for presence of aortic stenosis in adults (age 50 years or older). Am J Cardiol. 1985;55:744–7.

27. Aronow WS, Schwartz KS, Koenigsberg M. Correlation of serum lipids, calcium, and phosphorus, diabetes mellitus and history of systemic hypertension with presence or absence of calcified or thickened aortic cusps or root in elderly patients. Am J Cardiol. 1987;59:998–9.
28. Mohler ER, Sheridan MJ, Nichols R, Harvey WP, Waller BF. Development and progression of aortic valve stenosis: atherosclerosis risk factors—a causal relationship? A clinical morphologic study. Clin Cardiol. 1991;14:995–9.
29. Lindroos M, Kupari M, Valvanne J, Strandberg T, Heikkila J, Tilvis R. Factors associated with calcific aortic valve degeneration in the elderly. Eur Heart J. 1994;15:865–70.
30. Boon A, Cheriex E, Lodder J, Kessels F. Cardiac valve calcification: characteristics of patients with calcification of the mitral annulus or aortic valve. Heart. 1997;78:472–4.
31. Chui MC, Newby DE, Panarelli M, Bloomfield P, Boon NA. Association between calcific aortic stenosis and hypercholesterolemia: is there a need for a randomized controlled trial of cholesterol-lowering therapy? Clin Cardiol. 2001;24:52–5.
32. Wilmshurst PT, Stevenson RN, Griffiths H, Lord JR. A case-control investigation of the relation between hyperlipidaemia and calcific aortic valve stenosis. Heart. 1997;78:475–9.
33. Chan KL, Ghani M, Woodend K, Burwash IG. Case-controlled study to assess risk factors for aortic stenosis in congenitally bicuspid aortic valve. Am J Cardiol. 2001;88:690–3.
34. Briand M, Lemieux I, Dumesnil JG, et al. Metabolic syndrome negatively influences disease progression and prognosis in aortic stenosis. J Am Coll Cardiol. 2006;47:2229–36.
35. Palta S, Pai AM, Gill KS, Pai RG. New insights into the progression of aortic stenosis: implications for secondary prevention. Circulation. 2000;101:2497–502.
36. Peltier M, Trojette F, Sarano ME, Grigioni F, Slama MA, Tribouilloy CM. Relation between cardiovascular risk factors and nonrheumatic severe calcific aortic stenosis among patients with a three-cuspid aortic valve. Am J Cardiol. 2003;91:97–9.
37. Faggiano P, Antonini-Canterin F, Baldessin F, Lorusso R, D'Aloia A, Cas LD. Epidemiology and cardiovascular risk factors of aortic stenosis. Cardiovasc Ultrasound. 2006;4:27.
38. Pohle K, Maffert R, Ropers D, et al. Progression of aortic valve calcification: association with coronary atherosclerosis and cardiovascular risk factors. Circulation. 2001;104:1927–32.
39. Rosenhek R, Binder T, Porenta G, et al. Predictors of outcome in severe, asymptomatic aortic stenosis. N Engl J Med. 2000;343:611–7.
40. Messika-Zeitoun D, Aubry MC, Detaint D, et al. Evaluation and clinical implications of aortic valve calcification measured by electron-beam computed tomography. Circulation. 2004;110:356–62.
41. Bild DE, Detrano R, Peterson D, et al. Ethnic differences in coronary calcification: the multiethnic study of atherosclerosis (MESA). Circulation. 2005;111:1313–20.

2 Coronary Artery Calcification

Introduction

Coronary artery calcification (CAC) is a measure of the burden of atherosclerosis in the heart arteries and is measured by CT. The imaging technique measures the amount of calcification in the artery—the amount of bone, which develops in cholesterol-mediated atherosclerosis. Other components of the atherosclerotic plaque, including fatty (eg. cholesterol-rich components), often accompany CAC and can be present even in the absence of CAC—pre-clinical atherosclerosis—non-calcified.

Subclinical Atherosclerosis

Atherosclerosis is a chronic, progressive, inflammatory disease with a long asymptomatic phase, as shown in Fig. 2.1. This long asymptomatic phase of the disease mechanism, is the critical time point for identifying risk factors, initial stages of disease and any sign of early calcification to treat, modify and try to halt, slow or reverse progression. Disease progression can lead eventually to the occurrence of acute cardiovascular events such as myocardial infarction, unstable angina pectoris and sudden cardiac death. While the disease is still in a subclinical stage, however, the presence of atherosclerosis can be identified by several methods, including coronary angiography, intravascular ultrasonography, B-mode ultrasonography, computed tomography and magnetic resonance imaging. Based on the results of imaging studies, statin therapy can slow, halt or even reverse the progression of atherosclerotic disease, depending on the intensity of treatment. Whether to screen and treat patients for subclinical atherosclerosis remains controversial. Although atheroma plaque burden reduction has not yet been definitively correlated with significant decreases in risk for acute coronary events in asymptomatic patients, statin therapy contributes significantly to the risk reduction observed in clinical trials in patients with and

N.M. Rajamannan, *Osteocardiology*, DOI 10.1007/978-3-319-64994-8_2

Fig. 2.1 The progression of coronary artery atherosclerosis from subclinical to clinical disease

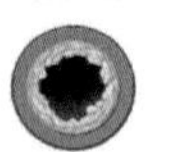

without overt coronary disease [1]. High dose statin therapy is not only associated with decreased cardiovascular events and mortality, in intravascular ultrasound studies high-dose statins are associated with and mild decrease of atheroma volume, particularly in women [2].

Some guidelines have recommended that screening for subclinical atherosclerosis, especially by CAC, might be appropriate in people at intermediate risk for heart disease (eg, 10-year estimate risk of 10–20%) but not for lower-risk general population screening or for people with preexisting heart disease or most other high–risk conditions. However, recent guidelines notes that those with diabetes mellitus who are ≥40 years of age may be suitable for screening or risk by coronary calcium [3]. According to the latest ACC/AHA cholesterol management guidelines [4], when treatment decisions are uncertain after 10-year risk is estimated, then the patient and clinician should take into consideration additional factors that modify the risk estimate, including an elevated CAC score or an ABI of >0.9. There are still limited data demonstrating whether screening with these and other imaging modalities can improve patient outcomes or whether it only increased downstream medical care costs. In part due to the complexity in defining the timing of therapy for the entire spectrum of the biologic effect of atherosclerosis—from subclinical to clinical manifestation of the disease, as shown in Fig. 2.1.

The Agatston Score

The Agatston score is a semi-automated tool to calculate a score based on the extent of coronary artery calcification detected by an unenhanced low-dose CT scan, which is routinely performed in patients undergoing cardiac CT. Due to an extensive body of research, it allows for an early risk stratification as patients with a high Agatston score (>160) have an increased risk for a major adverse cardiac event (MACE) [5].

The Agatston Score is the calculation of the amount of calcification, which is present in the anatomic location of the heart. The measurement of calcification in the heart with the use of CT imaging has helped to define the extent of calcification, and also to define prognosis in this patient population. The presence of any CAC, which indicates that at least some atherosclerotic plaque is present, is defined by an Agatston Score ≥100 or a score ≥75th percentile for one's age and sex; however, although they predict short-to intermediate-term risk, absolute CAC cutoffs offer more prognostic information across all age groups in both males and females. An Agatston score ≥400 has been noted to be an indication for further diagnostic evaluation (eg. exercise testing or myocardial perfusion imaging) for coronary artery disease (CAD). Further understanding of coronary calcification will help to understanding, the biology, risk to patients and future long-term clinical trials in this field to slow progression of disease [6].

Method of Calculation Agatston Score

The calculation is based on the weighted density score given to the highest attenuation value (HU) multiplied by area of the calcification speck. The calculation is based on the weighted density score given to the highest attenuation value (HU) multiplied by area of the calcification speck.

Density Factor

130–199 HU: 1
200–299 HU: 2
300–399 HU: 3
400+ HU: 4

For example, if a calcified spec has a maximum attenuation value of 400 HU and occupies 8 mm² area then its calcium score will be 32. The score of every calcified speck is summed up to give the total calcium score.

Grading of Coronary Artery Disease (Based on Total Calcium Score)

no evidence of CAD: 0 calcium score
minimal: 1–10
mild: 11–100
moderate: 101–400
severe: >400

Guidelines for Coronary Calcium Scoring by 2010 Task Force [3]

intermediate cardiovascular risk and asymptomatic adults (class IIa)
low-to-intermediate risk and asymptomatic adults (class IIb)
low risk and asymptomatic (class III)
asymptomatic adults with diabetes, 40 years of age and older (class IIa)

Prevalence of Coronary Artery Calcification

Coronary artery calcification (CAC) is highly prevalent in patients with coronary heart disease (CHD) and is associated with major adverse cardiovascular events. Further studies showed that the extent of coronary artery calcification (CAC) strongly correlated with the degree of atherosclerosis and the rate of future cardiac events [7]. The NHLBI's MESA [8] measured CAC in 6814 participants 45–84 years of age, including Caucasian (n = 2619), African American (n = 1898), Hispanic (n = 1494), and Chinese (n = 803) males and females. The prevalence and 75th percentile levels of CAC were highest in white males and lowest in African American and Hispanic females. Significant ethnic differences persisted after adjustment for risk factors, with the RR of coronary calcium being 225 less in African Americans, 15% less in Hispanics, and 8% less in Chinese than in whites.

In addition, a recent cost-effectiveness analysis based on data from MESA reported that CAC testing and statin treatment for those with CAC > 0 was cost effective in intermediate-risk scenarios (CV risk 5–10%) [9]. Furthermore, a recent MESA analysis compared these CAC-based treatment strategies to a "treat all" strategy and to treatment according to the ATPIII guidelines with clinical and economic modeled over both 5- and 10-year time horizons. The results consistently demonstrated that it is both cost-saving and more effective to scan intermediate–risk patients for CAC and to treat those with CAC ≥ 1 that to use treatment based on established risk assessment guidelines.

Renal failure is a unique subset of patients with more aggressive coronary artery calcification and coronary atherosclerosis. This is impacted by not only increased cardiovascular risk factors of hypertension and diabetes, but abnormalities in calcium and phosphorus metabolism contribute to intense calcific coronary disease seen. Coronary artery calcification even in the renal failure population continues to be a strong risk predictor for cardiovascular events [10]. This calcification in renal patients occurs not only in the intimal tissue as it does in non-renal patients but also occurs in the media, suggesting a unique mechanism of calcification in the renal failure population [11].

To date, effective medical treatment of CAC has not been identified. Several strategies of percutaneous coronary intervention have been applied to CHD patients with CAC, but with unsatisfactory results. Prognosis of CAC is still a major problem of CHD patients. Thus, more details about the mechanisms of CAC need to be elucidated in order to improve the understanding and treatment of CAC.

MESA Defines CAC Profiles

Coronary artery calcium (CAC) has been demonstrated to be associated with the risk of coronary heart disease. The Multi-Ethnic Study of Atherosclerosis (MESA) provides a unique opportunity to examine the distribution of CAC on the basis of age, gender, and race/ethnicity in a cohort free of clinical cardiovascular disease and treated diabetes. MESA is a prospective cohort study designed to investigate subclinical cardiovascular disease in a multiethnic cohort free of clinical cardiovascular disease. The percentiles of the CAC distribution were estimated with nonparametric techniques. Treated diabetics were excluded from analysis. There were 6110 included in the analysis, with 53% female and an average age of 62 years. Men had greater calcium levels than women, and calcium amount and prevalence were steadily higher with increasing age. There were significant differences in calcium by race, and these associations differed across age and gender. For women, whites had the highest percentiles and Hispanics generally had the lowest; in the oldest age group, however, Chinese women had the lowest values. Overall, Chinese and black women were intermediate, with their order dependent on age. For men, whites consistently had the highest percentiles, and Hispanics had the second highest. Blacks were lowest at the younger ages, and Chinese were lowest at the older ages. At the MESA public website (http://www.mesa-nhlbi.org), an interactive form allows one to enter an age, gender, race/ethnicity, and CAC score to obtain a corresponding estimated percentile. The information provided here can be used to examine whether a patient has a high CAC score relative to others with the same age, gender, and race/ethnicity who do not have clinical cardiovascular disease or treated diabetes [12].

MESA has defined the importance of measuring coronary artery calcium (CAC) in addition to traditional risk factors for coronary heart disease (CHD) risk prediction. This database is the first to developing a risk score incorporating CAC levels. In 2015, MESA developed a novel risk score to estimate 10-year CHD risk using CAC and traditional risk factors. Investigators developed algorithms in the MESA (Multi-Ethnic Study of Atherosclerosis), a prospective community-based cohort study of 6814 participants age 45–84 years, who were free of clinical heart disease at baseline and followed for 10 years. Inclusion of CAC in the MESA risk score offered significant improvements in risk prediction. Additionally, the difference in estimated 10-year risk between events and non-events was approximately 8–9%. Investigators determined that an accurate estimate of 10-year CHD risk can be obtained using traditional risk factors and CAC. The MESA risk score, which is available online on the MESA web site for easy use, can be used to aid clinicians when communicating risk to patients and when determining risk-based treatment strategies [13].

Coronary Artery Calcium (CAC) Score Reference Values web tool will provide the estimated probability of non-zero calcium, and the 25th, 50th, 75th, and 90th percentiles of the calcium score distribution for a particular age, gender and race. Additionally, if an observed calcium score is entered the program will provide the estimated percentile for this particular score. These reference values are based on participants in the MESA study who were free of clinical cardiovascular disease and treated diabetes at baseline. These participants were between 45 and 84 years of age, and identified themselves as White, African-American, Hispanic, or Chinese. The current tool is thus applicable only for these four race/ethnicity categories and within this age range. At this time, the risk associated with a particular calcium score is unknown. Thus, the information in this tool cannot necessarily be used to conclude that a patient is "high risk", but can indicate whether they have a high calcium score relative to others with the same age, gender, and race/ethnicity [13].

MESA in Coronary Artery Calcification Versus Calcific Aortic Valve Disease: The Role of Lp(a)

Of great mechanistic importance, the MESA dataset defined the role of Lp(a) in CAVD, but CAC was not associated with Lp(a). The inclusion of CAC into statistical models did not appreciably influence relations of Lp(a) and AVC in the sub-cohort or among races/ethnicities. The presence of existing coronary artery calcification did not affect these associations of Lp(a) and CAVD. There were no significant findings in Hispanics or Chinese. In contrast, CAC was only associated with CAVD in the sub-cohort using a regression model and adjusting for age, sex, education, diabetes, systolic blood pressure, hypertension meds, smoking, LDL, HDL, and triglycerides ($p < 0.001$). All of the traditional risk factors important in the development of CAVD [14].

MESA Defines Gene Expression Profiles

The MESA database also defined gene expression profiles by measuring RNA expression extracted from peripheral blood leukocytes. Coronary artery calcium (CAC) is a strong indicator of total atherosclerosis burden. Epidemiological data have shown substantial differences in CAC prevalence and severity between African Americans and whites. Microarray gene expression profiling of peripheral blood leucocytes was performed from 119 healthy women aged 50 years or above in the Multi-Ethnic Study of Atherosclerosis cohort; 48 women had CAC score >100 and carotid intima-media thickness (IMT) >1 mm, while 71 had CAC <10 and IMT <0.65 mm. When 17 African Americans were compared with 41 whites in the low-CAC group, 409 differentially expressed genes were identified. In addition, 316 differentially expressed genes were identified between the high- and low-CAC groups. Furthermore, genes expressed lower in African Americans also tend to express lower in individuals with low CAC. The data suggest a connection between immune response and vascular calcification and the result provides a potential mechanistic explanation for the lower prevalence and severity of CAC in African Americans compared with whites [15].

Furthermore, MESA demonstrated that low (<10%) to intermediate (10–20%) predicted Framingham risk; cases (N = 48) had coronary artery calcium (CAC) score > 100 and carotid intima-media thickness (IMT) >1.0 mm, whereas controls (N = 71) had CAC < 10 and IMT <0.65 mm. The RNA profiling study identified two major expression profiles significantly associated with significant atherosclerosis, among those with Framingham risk score <10%. Ontology analysis of the gene signature reveals activation of a major innate immune pathway, toll-like receptors and IL-1R signaling, in individuals with significant atherosclerosis. Gene expression profiles of peripheral blood may be a useful tool to identify individuals with significant burden of atherosclerosis, even among those with low predicted risk by clinical factors. Furthermore, the data suggest a critical association between atherosclerosis and the innate immune system and inflammation via TLR signaling in lower risk individuals [16].

Atherogenesis

Atherogenesis begins at sites of endothelial injury secondary to known risk factors for cardiovascular disease [3]. Early initiating events responsible for atherogenesis may result from a variety of factors, including increased local shear forces from hypertension, elevated plasma concentrations of LDL-C and remnant lipoprotein particles, cigarette smoke, low serum HDL-C and impaired reverse cholesterol transport, insulin resistance, and diabetes mellitus. These factors decrease endothelial cell production of nitric oxide, thereby impairing vasodilatory capacity and the normal barrier and protective functions of the vascular endothelium [17, 18]. As a result, LDL-C infiltrates the subendothelial space, where it can be oxidatively modified to initiate abnormal atherogenesis in the vessel.

Abnormal endothelial function, secondary to oxidative stress initiates a cascade of steps including attachment of circulation monocytes, activation of growth factors and development of an atherosclerotic lesion, compose of lipid-laden foam cells, proliferating and synthetic vascular smooth muscle cells which are differentiating into an osteogenic phenotype for future calcification, or bone formation.

The Molecular Biology of Vascular Calcification

The role of lipids in vascular calcification, have been the focus of intense investigation over the past 100 years. Lipids and other cardiovascular risk factors induce oxidative stress [19–21] in the aortic valve endothelium similar to vascular endothelium [22] which in turn activates the secretion of cytokines and growth factors important in cell signaling as shown in Fig. 2.2. The early atherosclerotic and abnormal oxidative stress environment also plays a role in the activation of the calcification process in the myofibroblast cell. Cardiovascular risk factors, cell proliferation [23] and cyclic stretch [24] play a role in the activation of these cells to transition to a calcifying phenotype. There is also increasing evidence that these cells undergo specific differentiation steps towards the development of this bone phenotype as shown in *in vitro* studies [25–27]. The signaling molecules important in the development of vascular atherosclerosis are also important in the development of valve calcification including: MMP [28, 29], Interleukin 1 [30], transforming growth factor-beta (TGF-beta) [31], purine nucleotides [32, 33], RANK [34], osteoprotegrin (OPG) [34], elastolytic cathepsins S, K, and V and their inhibitor Cystatin C in stenotic aortic valves [35] Toll-like receptors [36],TNF alpha [37], MAP Kinase [23] and the canonical Wnt pathway [38–40]. Similar to vascular atherosclerosis these events are potential cellular targets for pharmacologic agents to slow this disease process. HMG CoA Reductase agents, angiotensin converting enzyme (ACE)

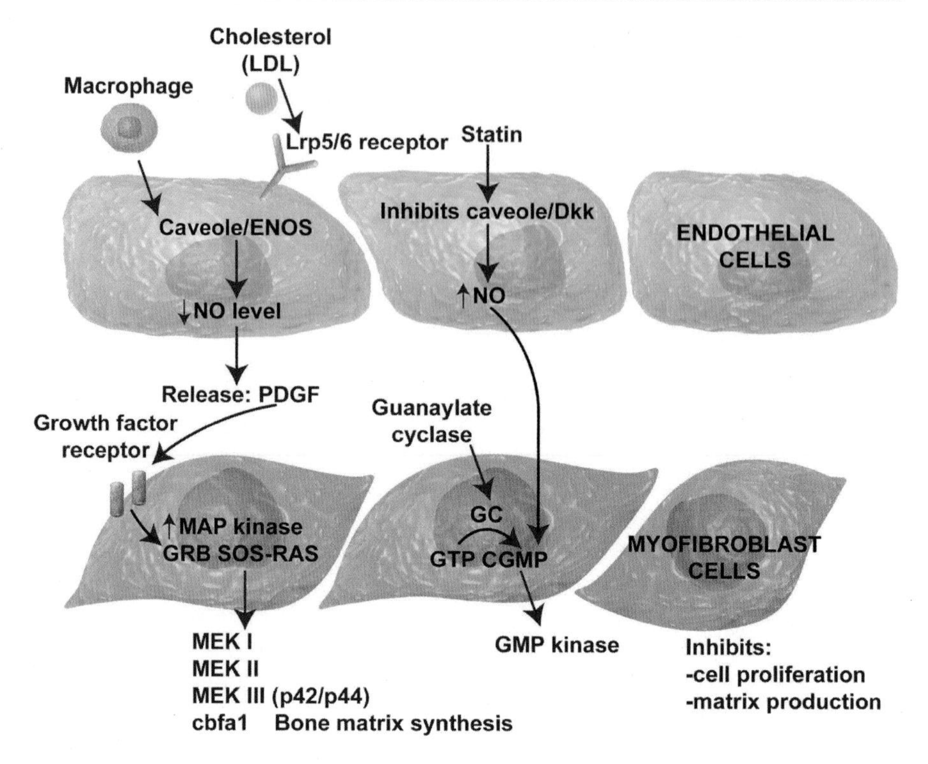

Fig. 2.2 Signaling pathways in the development of vascular atherosclerosis

inhibitors, and angiotensin receptor blockers (ARBs), provide an interesting approach for targeting in this disease.

Interventional Therapy for CAC

CAC increases the likelihood of procedural failure and complications after balloon angioplasty [41]. Besides, the force applied from the balloon to the vessel wall might not be uniform across the length of the lesion, due to varying amounts of calcification, which increases the risk for dissection and acute vessel closure, MI, restenosis, and MACE [42]. Rotational atherectomy abrades hard tissue into smaller particles (<10 μm) while deflecting off softer elastic tissue [43]. Therefore, rotational atherectomy has a selective effect on hard lesions, but not the soft tissues. In the pre-stent era, the use of rotational atherectomy alone was associated with increased neo-intimal hyperplasia, restenosis, and repeat revascularization, which was most likely due to platelet activation and thermal injury [44]. Excimer laser coronary atherectomy (ELCA) can dilate resistant lesions through a photoacoustic mechanism. In-stent restenosis can be treated by ELCA with similar outcomes as rotational atherectomy [45].

Similarly, it can facilitate stent expansion when high-pressure non-compliant balloon inflation fails to adequate expand a stent due to calcific or fibrotic coronary disease.

Orbital atherectomy is a newer form of atherectomy, which utilizes an orbiting eccentric diamond-coated crown which removes plaque by creating increasing debulked areas at the tip by an increasing size of an orbital field as the speed of the device is increased. (Diamondback 360° Orbital Atherectomy System, Cardiovascular Systems, Inc., St. Paul, MN). In ORBIT II this device was shown to facilitate stent delivery and improved outcomes compared with historic controls [46]. While feasible, the specific role of orbital atherectomy in percutaneous coronary revascularization awaits prospective randomized trials to demonstrate benefit.

Coronary artery bypass graft surgery remains a viable option for revascularization and has an increased role in patients with class III to IV CAC based on intravascular ultrasound, as well as anticipated difficulty in performing percutaneous coronary revascularization. However, increased morbidity and mortality occurs from challenges in bypassing the coronary artery and increased embolic complications with cross-clamping a calcified and atherosclerotic aorta [47, 48].

Thus, coronary artery calcification remains not only a marker of early atherosclerosis it is an end-stage manifestation of severe coronary atherosclerosis which challenges our ability for safe and successful revascularization. Determining how to balance the beneficial effects of calcification on plaque stability while gaining incite on how to safely influence this process and potentially reverse it to facilitate revascularization will be needed to advance the treatment of coronary artery disease.

Summary

Results from MESA [49] and the current guidelines of the treatment of cardiovascular heart disease [3], have indicated that CAC screening is most useful for identifying patients with early atherosclerosis, with the most powerful identifier is the patient with no CAC, who do not need therapy. The absence of subclinical atherosclerosis, indicates that patient who are at low risk have a better long-term survival [49].

References

1. Weintraub WS, Pederson JP. Atherosclerosis and restenosis: reflections on the Lovastatin Restenosis Trial and Scandinavian Simvastatin Survival Study. Am J Cardiol. 1996;78:1036–8.
2. Nicholls SJ, Ballantyne CM, Barter PJ, et al. Effect of two intensive statin regimens on progression of coronary disease. N Engl J Med. 2011;365:2078–87.
3. Greenland P, Alpert JS, Beller GA, et al. 2010 ACCF/AHA guideline for assessment of cardiovascular risk in asymptomatic adults: a report of the American College of Cardiology Foundation/American Heart Association Task Force on Practice Guidelines. J Am Coll Cardiol. 2010;56:e50–103.

4. Stone NJ, Robinson J, Lichtenstein AH, et al. ACC/AHA guideline on the treatment of blood cholesterol to reduce atherosclerotic cardiovascular risk in adults: a report of the American College of Cardiology/American Heart Association Task Force on Practice Guidelines. Circulation. 2013;2013. [Epub ahead of print]
5. van der Bijl N, Joemai RM, Geleijns J, et al. Assessment of Agatston coronary artery calcium score using contrast-enhanced CT coronary angiography. AJR Am J Roentgenol. 2010;195:1299–305.
6. Benjamin EJ, Blaha MJ, Chiuve SE, et al. Heart disease and stroke statistics-2017 update: a report from the American Heart Association. Circulation. 2017;135:e146–603.
7. Greenland P, LaBree L, Azen SP, Doherty TM, Detrano RC. Coronary artery calcium score combined with Framingham score for risk prediction in asymptomatic individuals. JAMA. 2004;291:210–5.
8. Detrano R, Guerci AD, Carr JJ, et al. Coronary calcium as a predictor of coronary events in four racial or ethnic groups. N Engl J Med. 2008;358:1336–45.
9. Pletcher MJ, Pignone M, Earnshaw S, et al. Using the coronary artery calcium score to guide statin therapy: a cost-effectiveness analysis. Circ Cardiovasc Qual Outcomes. 2014;7:276–84.
10. Goodman WG, Goldin J, Kuizon BD, et al. Coronary-artery calcification in young adults with end-stage renal disease who are undergoing dialysis. N Engl J Med. 2000;342:1478–83.
11. Nakamura S, Ishibashi-Ueda H, Niizuma S, Yoshihara F, Horio T, Kawano Y. Coronary calcification in patients with chronic kidney disease and coronary artery disease. Clin J Am Soc Nephrol. 2009;4:1892–900.
12. McClelland RL, Chung H, Detrano R, Post W, Kronmal RA. Distribution of coronary artery calcium by race, gender, and age: results from the Multi-Ethnic Study of Atherosclerosis (MESA). Circulation. 2006;113:30–7.
13. McClelland RL, Jorgensen NW, Budoff M, et al. 10-year coronary heart disease risk prediction using coronary artery calcium and traditional risk factors: derivation in the MESA (Multi-Ethnic Study of Atherosclerosis) with validation in the HNR (Heinz Nixdorf Recall) Study and the DHS (Dallas Heart Study). J Am Coll Cardiol. 2015;66:1643–53.
14. Rajamannan NM, Evans FJ, Aikawa E, et al. Calcific aortic valve disease: not simply a degenerative process: a review and agenda for research from the National Heart and Lung and Blood Institute Aortic Stenosis Working Group. Executive summary: calcific aortic valve disease-2011 update. Circulation. 2011;124:1783–91.
15. Huang CC, Lloyd-Jones DM, Guo X, et al. Gene expression variation between African Americans and whites is associated with coronary artery calcification: the multiethnic study of atherosclerosis. Physiol Genomics. 2011;43:836–43.
16. Huang CC, Liu K, Pope RM, et al. Activated TLR signaling in atherosclerosis among women with lower Framingham risk score: the multi-ethnic study of atherosclerosis. PLoS One. 2011;6:e21067.
17. Best PJ, McKenna CJ, Hasdai D, Holmes DR Jr, Lerman A. Chronic endothelin receptor antagonism preserves coronary endothelial function in experimental hypercholesterolemia. Circulation. 1999;99:1747–52.
18. Best PJ, Lerman LO, Romero JC, Richardson D, Holmes DR Jr, Lerman A. Coronary endothelial function is preserved with chronic endothelin receptor antagonism in experimental hypercholesterolemia in vitro. Arterioscler Thromb Vasc Biol. 1999;19:2769–75.
19. Rajamannan NM, Subramaniam M, Stock SR, et al. Atorvastatin inhibits calcification and enhances nitric oxide synthase production in the hypercholesterolaemic aortic valve. Heart. 2005;91:806–10.
20. Weiss RM, Ohashi M, Miller JD, Young SG, Heistad DD. Calcific aortic valve stenosis in old hypercholesterolemic mice. Circulation. 2006;114:2065–9.
21. Miller JD, Chu Y, Brooks RM, Richenbacher WE, Pena-Silva R, Heistad DD. Dysregulation of antioxidant mechanisms contributes to increased oxidative stress in calcific aortic valvular stenosis in humans. J Am Coll Cardiol. 2008;52:843–50.
22. Wilcox JN, Subramanian RR, Sundell CL, et al. Expression of multiple isoforms of nitric oxide synthase in normal and atherosclerotic vessels. Arterioscler Thromb Vasc Biol. 1997;17:2479–88.

23. Gu X, Masters KS. Role of the MAPK/ERK pathway in valvular interstitial cell calcification. Am J Physiol. 2009;296:H1748–57.
24. Balachandran K, Sucosky P, Jo H, Yoganathan AP. Elevated cyclic stretch alters matrix remodeling in aortic valve cusps: implications for degenerative aortic valve disease. Am J Physiol. 2009;296:H756–64.
25. Blevins TL, Peterson SB, Lee EL, et al. Mitral valvular interstitial cells demonstrate regional, adhesional, and synthetic heterogeneity. Cells Tissues Organs. 2008;187:113–22.
26. Liu AC, Joag VR, Gotlieb AI. The emerging role of valve interstitial cell phenotypes in regulating heart valve pathobiology. Am J Pathol. 2007;171:1407–18.
27. Yip CY, Chen JH, Zhao R, Simmons CA. Calcification by valve interstitial cells is regulated by the stiffness of the extracellular matrix. Arterioscler Thromb Vasc Biol. 2009;29: 936–42.
28. Kaden JJ, Vocke DC, Fischer CS, et al. Expression and activity of matrix metalloproteinase-2 in calcific aortic stenosis. Z Kardiol. 2004;93:124–30.
29. Jian B, Jones PL, Li Q, Mohler ER 3rd, Schoen FJ, Levy RJ. Matrix metalloproteinase-2 is associated with tenascin-C in calcific aortic stenosis. Am J Pathol. 2001;159:321–7.
30. Kaden JJ, Dempfle CE, Grobholz R, et al. Interleukin-1 beta promotes matrix metalloproteinase expression and cell proliferation in calcific aortic valve stenosis. Atherosclerosis. 2003;170:205–11.
31. Jian B, Narula N, Li QY, Mohler ER 3rd, Levy RJ. Progression of aortic valve stenosis: TGF-beta1 is present in calcified aortic valve cusps and promotes aortic valve interstitial cell calcification via apoptosis. Ann Thorac Surg. 2003;75:457–65. discussion 65–6
32. Osman L, Chester AH, Amrani M, Yacoub MH, Smolenski RT. A novel role of extracellular nucleotides in valve calcification: a potential target for atorvastatin. Circulation. 2006;114:I566–72.
33. Osman L, Amrani M, Isley C, Yacoub MH, Smolenski RT. Stimulatory effects of atorvastatin on extracellular nucleotide degradation in human endothelial cells. Nucleosides Nucleotides Nucleic Acids. 2006;25:1125–8.
34. Kaden JJ, Bickelhaupt S, Grobholz R, et al. Receptor activator of nuclear factor kappaB ligand and osteoprotegerin regulate aortic valve calcification. J Mol Cell Cardiol. 2004;36:57–66.
35. Helske S, Syvaranta S, Lindstedt KA, et al. Increased expression of elastolytic cathepsins S, K, and V and their inhibitor cystatin C in stenotic aortic valves. Arterioscler Thromb Vasc Biol. 2006;26:1791–8.
36. Yang X, Fullerton DA, Su X, Ao L, Cleveland JC Jr, Meng X. Pro-osteogenic phenotype of human aortic valve interstitial cells is associated with higher levels of Toll-like receptors 2 and 4 and enhanced expression of bone morphogenetic protein 2. J Am Coll Cardiol. 2009;53:491–500.
37. Kaden JJ, Kilic R, Sarikoc A, et al. Tumor necrosis factor alpha promotes an osteoblast-like phenotype in human aortic valve myofibroblasts: a potential regulatory mechanism of valvular calcification. Int J Mol Med. 2005;16:869–72.
38. Shao JS, Cheng SL, Pingsterhaus JM, Charlton-Kachigian N, Loewy AP, Towler DA. Msx2 promotes cardiovascular calcification by activating paracrine Wnt signals. J Clin Invest. 2005;115:1210–20.
39. Rajamannan NM, Subramaniam M, Caira F, Stock SR, Spelsberg TC. Atorvastatin inhibits hypercholesterolemia-induced calcification in the aortic valves via the Lrp5 receptor pathway. Circulation. 2005;112:I229–34.
40. Caira FC, Stock SR, Gleason TG, et al. Human degenerative valve disease is associated with up-regulation of low-density lipoprotein receptor-related protein 5 receptor-mediated bone formation. J Am Coll Cardiol. 2006;47:1707–12.
41. Bourantas CV, Zhang YJ, Garg S, et al. Prognostic implications of coronary calcification in patients with obstructive coronary artery disease treated by percutaneous coronary intervention: a patient-level pooled analysis of 7 contemporary stent trials. Heart. 2014;100:1158–64.
42. Fitzgerald PJ, Ports TA, Yock PG. Contribution of localized calcium deposits to dissection after angioplasty. An observational study using intravascular ultrasound. Circulation. 1992;86:64–70.

43. Zimarino M, Corcos T, Bramucci E, Tamburino C. Rotational atherectomy: a "survivor" in the drug-eluting stent era. Cardiovasc Revasc Med. 2012;13:185–92.
44. MacIsaac AI, Bass TA, Buchbinder M, et al. High speed rotational atherectomy: outcome in calcified and noncalcified coronary artery lesions. J Am Coll Cardiol. 1995;26:731–6.
45. Mehran R, Dangas G, Mintz GS, et al. Treatment of in-stent restenosis with excimer laser coronary angioplasty versus rotational atherectomy: comparative mechanisms and results. Circulation. 2000;101:2484–9.
46. Chambers JW, Feldman RL, Himmelstein SI, et al. Pivotal trial to evaluate the safety and efficacy of the orbital atherectomy system in treating de novo, severely calcified coronary lesions (ORBIT II). JACC Cardiovasc Interv. 2014;7:510–8.
47. Castagna MT, Mintz GS, Ohlmann P, et al. Incidence, location, magnitude, and clinical correlates of saphenous vein graft calcification: an intravascular ultrasound and angiographic study. Circulation. 2005;111:1148–52.
48. Roach GW, Kanchuger M, Mangano CM, et al. Adverse cerebral outcomes after coronary bypass surgery. Multicenter Study of Perioperative Ischemia Research Group and the Ischemia Research and Education Foundation Investigators. N Engl J Med. 1996;335:1857–63.
49. Budoff MJ, Nasir K, McClelland RL, et al. Coronary calcium predicts events better with absolute calcium scores than age-sex-race/ethnicity percentiles: MESA (Multi-Ethnic Study of Atherosclerosis). J Am Coll Cardiol. 2009;53:345–52.

Osteocardiology: Calcific Aortic Valve Disease

3

Introduction

Calcific aortic valve disease is the most common indication for valve intervention in the world [1]. The cellular mechanisms, cardiovascular risk factors and therapeutic interventions have been under intense investigation in the twenty-first Century. Calcific aortic valve disease (CAVD) covers a spectrum of disease from early initiation stages, through cell differentiation, cell proliferation and extracellular matrix production, causing aortic sclerosis, progressive outflow obstruction and eventual decreased leaflet mobility and aortic stenosis. The different changes in the evolution of disease has been characterized by the advent of echocardiography, although for decades, clinical auscultation has provided an excellent approach to diagnosing severity of disease, from sclerosis to stenosis.

The key features of auscultation of CAVD include the classic auscultation finding of an early systolic ejection click with a bicuspid aortic valve. Mild aortic valve sclerosis is associated with a soft early systolic murmer. Moderate aortic valve stenosis is associated with a loud 2–3/6 mid systolic murmer. Severe aortic stenosis is associated with a late peaking systolic murmer, associated with Parvus and Tardus of the carotid pulse. The ACC/AHA 2014 Valvular Heart Disease guidelines classify aortic valve disease according to echocardiography hemodynamics using a four stage grading classification: STAGE A: At risk of AS, bicuspid aortic valve, Aortic Velocity <2 m/s, with mild aortic valve sclerosis; STAGE B: Progressive AS, Mild AS, Aortic Velocity 2–2.9 m/s or a change in pressure <20 mmHg, Moderate AS, Aortic Velocity 3.0–3.9 m/s, or a mean change in pressure 20–39 mmHg, with associated mild-moderate leaflet calcification; STAGE C: Asymptomatic Severe AS, Aortic Velocity ≥4 m/s or a mean change in Pressure ≥40 mmHg, AVA typically is ≤1.0 cm^2, Very Severe AS is an aortic Velocity ≥5 m/s or a mean change in pressure ≥60 mmHg, associated with severe leaflet calcification; STAGE D: Symptomatic severe AS, Aortic velocity ≥4 m/s or a mean change in pressure ≥40 mmHg, AVA typically ≤1.0 cm^2, with associated severe leaflet calcification [2].

N.M. Rajamannan, *Osteocardiology*, DOI 10.1007/978-3-319-64994-8_3

The later stages of severe calcific aortic valve disease are characterized by calcific thickening of the valve leaflets and the formation of neoangiogenesis [3] and calcium nodules—often including the formation of actual bone [4]—throughout the valve leaflets [5] but concentrated near the aortic surface of the valve and not the ventricular surface. For decades, CAVD was thought to be dues to a degenerative process, but now CAVD is an actively regulated disease process that cannot be characterized simply as "senile" or "degenerative." anymore [1].

Epidemiological studies show that some of the risk factors for CAVD are similar to those for vascular atherosclerosis [6] and osteoporosis [7]. Age, gender, hypertension, diabetes mellitus, gender, end-stage renal disease, are just a few of the associated with an increased risk of CAVD. The actual steps in the development of CAVD are still under intense investigation. With the emerging animal models, the stages of the cellular biology of CAVD will soon be elucidates in the research laboratory. CAVD may progress to a point-of-no-return, when no medical therapy will slow the progression of disease [8]. Whether a point-of-no-return or a "no go" stage really exists, and if so, whether it's a fundamental aspect of CAVD biology- or the steps in the osteogenic cascade of bone formation are critical towards further elucidating these time points.

Three Stages of Calcific Aortic Valve Disease

The histopathologic, epidemiologic, imaging, and experimental mechanistic studies have defined the three stages of CAVD, which include: The normal aortic valve, mild to moderate aortic valve stenosis, of the subclinical stage, and severe asymptomatic to symptomatic aortic stenosis, or the clinical stage. Figure 3.1, demonstrates the three stages and which will provide the reference point for this textbook—osteocardiology.

The Role of CHS and MESA in Calcific Aortic Valve Disease (CAVD)

Calcific aortic valve disease is estimated to have a prevalence of 25% in individuals over 65 years of age [6]. Thought previously to be a degenerative disorder, the disease now is recognized to be an actively-regulated biological process sharing many epidemiologic [1], and histopathologic [1], similarities to coronary atherosclerosis.

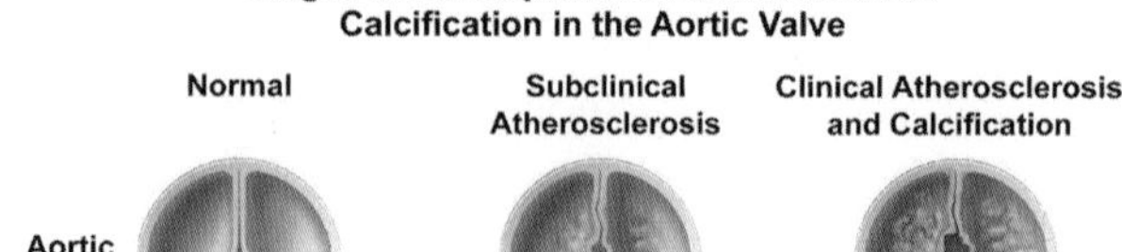

Fig. 3.1 MESA defines subclinical versus clinical disease as measured by valve calcium scores, and associations with traditional cardiovascular risk factors

In patients without known cardiovascular disease, aortic sclerosis (the presence of valve calcium without hemodynamic obstruction) from the Cardiovascular Health Study (CHS) showed that, among adults >65 years, echocardiographic detection of aortic valve sclerosis was associated with a 50% increased risk of cardiovascular mortality [9]. In that study, aortic sclerosis also was associated with a 42% increase in risk of MI [9]. However, these analyses were unable to control for the presence of subclinical atherosclerosis and systemic inflammation, plausible mediators of these associations.

It is well known that calcific aortic valve disease begins in midlife as a clinically latent but progressive disorder, and often is detected incidentally. Yet even in this latent, pre-obstructive phase, the presence of "aortic valve calcium" appears to be a marker of increased cardiovascular risk. However, these analyses were unable to control for the presence of subclinical atherosclerosis and systemic inflammation, plausible mediators of these associations.

To determine whether the presence of aortic valve calcium (AVC), detected from computed tomography scans, predicts cardiovascular events in a younger cohort, and to identify mechanisms underlying this association, the MESA database provides the foundation to determine the role of subclinical risk factors in younger people who do not have overt cardiovascular disease. MESA researchers performed several prospective analysis of the MESA cohort, as described in Chap. 1.

In 2010, MESA defined the association of subclinical risk factors and new diagnosis of CAVD and or it progression [10]. This study may be the most important in terms of diagnosing aortic valve disease in patients without symptoms. In the study, CAVD was quantified from serial computed tomographic images from 5880 participants (aged 45–84 years) in the Multi-Ethnic Study of Atherosclerosis cohort, using the Agatston method. During a mean follow-up of 2.4 ± 0.9 years, 210 subjects (4.1%) developed incident CAVD. The incidence rate (mean 1.7%/year) increased significantly with age ($p < 0.001$). The risk factors in MESA associated with the newly diagnosed CAVD, included age, male gender, body mass index, current smoking, and the use of lipid-lowering and antihypertensive medications. Among those with CAVD at baseline, the median rate calcification progression was 2 Agatston units/year [10]. The baseline Agatston score was a strong, independent predictor of progression, especially among those with high calcium scores at baseline. In conclusion, in this MESA, preclinical cohort, the rate of incident CAVD increased significantly with age. The incident CAVD risk was associated with several traditional cardiovascular risk factors, specifically age, male gender, body mass index, current smoking, and the use of both antihypertensive and lipid-lowering medications. CAVD progression risk was associated with male gender and the baseline Agatston score. Additional research is needed to determine whether age- and stage-specific mechanisms underlie the risk of CAVD progression.

The same group of investigators found in the MESA cohort, free of clinical cardiovascular disease, that CAVD predicts cardiovascular and coronary event risk independent of traditional risk factors and inflammatory biomarkers, likely due to the strong correlation between CAVD and subclinical atherosclerosis [11]. Importantly, they found that aortic valve calcium independently predicts coronary and cardiovascular events in a primary prevention MESA population [11]. MESA is instrumental in defining the concept that calcific aortic valve disease begins in midlife as a clinically asymptomatic, but progressive disorder, and normally diagnosed as an incidental finding. Yet even in the long latent, pre-stenotic phase, the presence of aortic valve calcium appears to be a marker of increased cardiovascular risk.

Lipoproteins as Novel Risk Factors in Calcific Aortic Valve Disease

Otto and O'Brien, are the first to publish studies to define the role of lipoproteins in *ex vivo* calcified aortic valves [12, 13]. Over the next 20 years, studies in the field of calcific aortic valve disease have determined that the calcific aortic valve disease is not a degenerative process, but an active cellular biology [1]. This hypothesis was confirmed using animal models, which tested the role of hypercholesterolemia as an initiating event for calcific aortic valve disease [14]. In 2009, NHLBI convened a working group on the cellular mechanisms of calcific aortic valve disease [1]. The working hypothesis for the development of calcific aortic valve disease emphasized the role of lipoproteins and oxidative stress in the initiation of the disease [15]. Calcification ensues over time as aortic valve myofibroblasts differentiate into an osteogenic phenotype [4, 5, 16].

Lp(a) in Calcific Aortic Valve Disease

In 2013, Thanassoulis et al. [17] studied role of common genetic variation in valvular calcification. Genetic determinants of valvular calcification may help elucidate the mechanisms underlying valvular heart disease, and could identify new therapies. The investigators performed a genome-wide association study of aortic-valve calcification and mitral annular calcification in three population-based cohorts. The results were confirmed in additional multiethnic cohorts by means of computed tomographic (CT) assessment of valvular calcification or identification of clinically apparent valvular heart disease.

The investigation was initiated within the Cohorts for Heart and Aging Research in Genome Epidemiology (CHARGE) consortium. They then performed a two-stage analysis to discover the associations of genetic loci with the presence of mitral annular calcification and aortic-valve calcification and to confirm the findings in the first cohort during the replication phase of the study the investigators used several databases including the FHS, the MESA database. The findings discovered the role of genetic variation in the *LPA* locus, mediated by Lp(a) levels, is associated with aortic-valve calcification across multiple ethnic groups and with incident clinical aortic stenosis. [17]

Thanassoulis has further proposed that targeted therapy of Lp(a) may be a novel target for treating calcific aortic valve disease, after confirming the genetics of Lp(a) in patients with CAVD [18, 19]. MESA also confirmed the discovery of Lp(a) as a significant risk factor for CAVD [20]. MESA was designed to test subclinical atherosclerosis markers, and measure calcification burden in the aortic valve using Computed Tomography (CT) measurements. The study group included individuals from age 45 to 84, who were free of any clinical cardiovascular disease and treated diabetes [21, 22].

Cao et al. [20], confirmed the role of Lp(a) in MESA as a risk factor or CAVD while sorting out the role of other traditional risk factors versus Lp(a). The MESA dataset, and the inclusion of CAC did not appreciably influence relations of Lp(a) and CAVD in the subcohort or among races/ethnicities. MESA has played an important role in studying the development of calcification and defining subclinical risk factors in aortic valve calcification and coronary artery calcification [10, 23–25]. In a recent study [20], investigations define the cut-offs for Lp(a) and the effect of ethnicity in the development of CAVD. Lp(a) concentrations were measured using a turbidimetric immunoassay, and subclinical CAVD was measured by quantifying calcific aortic valve disease (CAVD) through computed tomography scanning in 4678 participants of the Multi-Ethnic Study of Atherosclerosis. Relative risk (RR) and ordered logistic regression analysis determined cross-sectional associations of Lp(a) with CAVD and its severity, respectively. The conventional 30 mg/dL Lp(a) clinical cut-off was associated with CAVD in Caucasian and was borderline significant ($p = 0.059$) in African American study participants. Caucasians with levels ≥50 mg/dL also showed higher prevalence of CAVD than those below this level. Significant associations were observed between Lp(a) and degree of CAVD in both Caucasians and African American individuals. The degree of CAVD in Asians and Hispanic is not significant, but this could be due to the results being underpowered in these populations tested. The finding in the aortic valve differentiates the role of Lp(a) in the progression of aortic valve calcification, which is not related to the presence of CAC in the population. This may be due to the mechanism of coronary artery calcification [26] versus aortic valve calcification [4], versus the role of embryonic cell linage [27] in the mechanism of calcification. Previous investigations have also defined the importance of Lp(a) in the development of CAVD in genetic studies [19].

The role of multiple lipoproteins in the progression of CAVD, may further aid in understanding the outcomes of the clinical trials in CAVD [8, 28–30] designed to lower lipids in calcific aortic valve disease. ASTRONOMER specifically addressed this question by determining the role of oxidized phospholipids and Lp(a) in the progression of calcific aortic valve disease [31]. It is well known that traditional risk factors play a role in the majority of patients with CAVD [1], however, genetics [32] and lipoprotein Lp(a) [20] are also critical in the development of CAVD. Figure 3.2, demonstrates cardiovascular risk factors, genetic factors, and the final calcification phenotype critical in the development of calcific aortic valve disease. Future studies evaluating the role of Lp(a) in patients in with CAVD may help to further understand the role of lipoprotein driving early atherosclerosis and eventual calcification in the aortic valve.

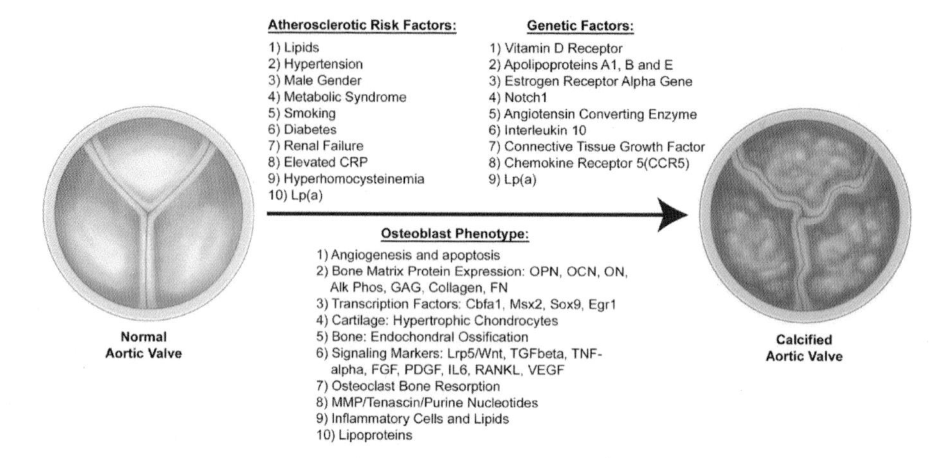

Fig. 3.2 This figure describes the role of atherosclerotic risk factors published to date in the field of CAVD, traditional and non-traditional risk factors (Lp(a)), the role of genetics and defining the signaling pathways upregulated in the final common pathway in the development of calcific aortic valve disease [81] (Permission obtained to reproduce the figure)

Bicuspid Aortic Valve Disease

Although aortic stenosis may occur in individuals with otherwise anatomically normal aortic valves, congenital valve abnormalities markedly increase the risk. Nearly half of the individuals with aortic stenosis have a bicuspid aortic valve (BAV) [33], an aortic valve that developed with two functional leaflets instead of the normal three. BAV occurs in about 0.6% of the population and is the most common congenital cardiac malformation. Although the causes of BAV are unclear, genetic factors have been identified in some cases. CAVD tends to develop at an earlier age in individuals with BAV and to progress more rapidly for reasons that have been poorly understood until recent in vivo animal models have clearly demonstrate the rapid progression of the BAV versus TAV, in hyperlipidemic eNOS null mouse [15]. Furthermore, studies have shown that the eNOS$^{-/-}$ mouse is a novel mouse model, which develops anatomic bicuspid aortic valves (BAV) in approximately 25% of the eNOS null mouse population [15, 34]. Genetic mutations associated with BAV that cause cellular dysfunction may also predispose an individual to other congenital heart defects or to dilation and dissection of the ascending aorta.

Mitral Annular Calcification (MAC)

Several similarities exist between atherosclerosis in the vasculature and chronic degenerative changes in valvular structures. It has been suggested that aortic valve sclerosis (AVS) and mitral annulus calcification (MAC) are manifestations of a generalized atherosclerosis, have similar pathogenesis, share common risk factors and are observed with higher prevalence in patients with different forms of atherosclerotic vascular disease including carotid artery disease, coronary artery disease, and aortic atheroma. Moreover, recent studies have shown a close relation of MAC and CAVD with adverse cardiovascular and cerebrovascular outcomes. However, many patients with CAVD or MAC do not have coexisting peripheral vascular atherosclerosis and vice versa. Thus, whether valve calcifications are the result of a more generalized atherosclerosis, or reflect a primary degenerative process, progressing with advancing age, still remains. From a clinical point of view it is of great importance to identify common links between valve calcification and vascular atherosclerosis with a view to assess whether the detection of AVS, MAC or both is indicative of subclinical atherosclerosis and predicts cardiovascular or cerebrovascular events [35].

Overview of the NHLBI Working Group Consensus Panel

Normal Aortic Valve Function and Anatomy

The normal function of the heart valve is to permit unidirectional forward flow through the cardiovascular circulation. The valve components must accomplish the second-to-second movements necessitated by the cardiac cycle and must maintain strength and durability to withstand repetitive mechanical stress and strain over many years. The movement through the cardiac cycle, and the ability to endure the stress imposed on the valve over the lifetime is accomplished by a specific cellular architecture [36].

The aortic valve (AV) as viewed by echocardiography and bioreactor models (Fig. 3.3, Panel a). The direction of flow during systole is allowing the valve cusps to open as the blood flows across the open aortic valve leaflets. The inflow surface is the located along the direction of flow as indicated in Fig. 3.3, Panel a. The outflow surface is demonstrated in the diastole figure as the valves are closed and there is end diastolic pressure closing the valve leaflets along the outflow surface. During diastole, the tissue of the cusps is stretched via a backpressure; during systole, the cusp tissue becomes relaxed and shortens owing to recoil of elastin, which was elongated during diastole.

All four cardiac valves have a similar layered architectural pattern composed of cells, including the valvular endothelial cells, the deep valvular interstitial cell (VICs), and valvular extracellular matrix, including collagen, elastin and glycosaminoglycans. The outflow surface provides strength: the *fibrosa*; a central core of loose connective tissue: the *spongiosa* rich in glycosaminoglycans (GAGs); and a layer rich in elastin below the inflow surface: the *ventricularis* as shown in (Fig. 3.3, Panel b). Research has developed specific culture techniques to isolate interstitial cells, which have been used to demonstrate that discrete valvular cell subpopulations have unique morphological characteristics, synthesis of ECM, potential for calcification and ossification, and potential for promoting angiogenesis [3]. All characteristics important in the development of calcific valve disease, and the development of the progression of disease, which is the calcified valve leaflet as depicted in (Fig. 3.3, Panel c).

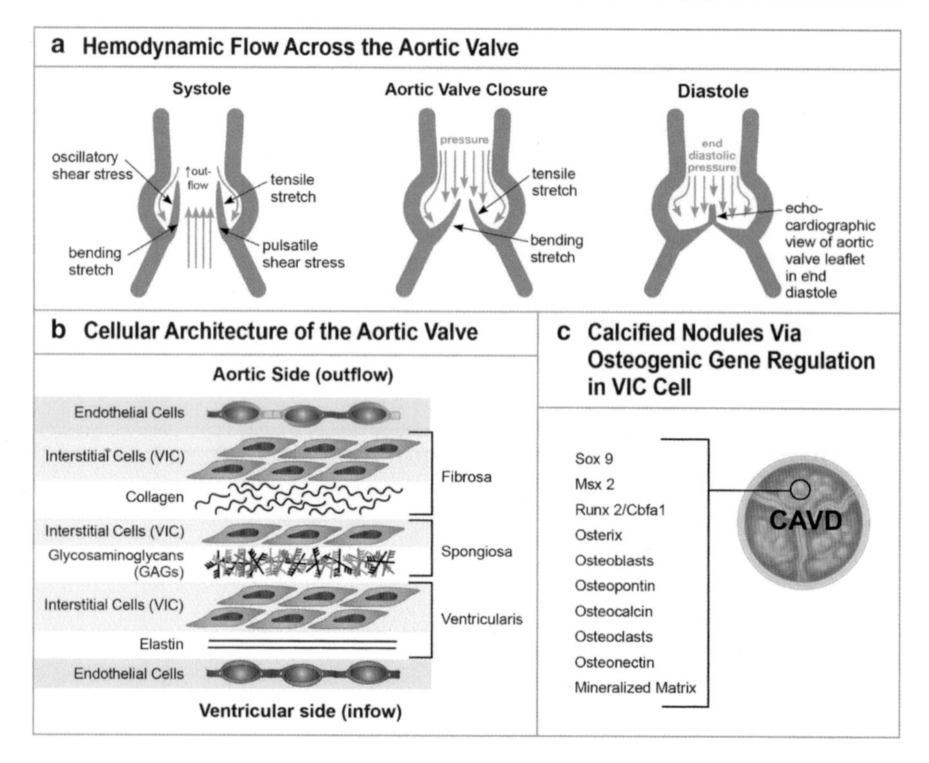

Fig. 3.3 Echocardiographic and Bioengineering and Hemodynamic Force Perspective of the diastole and systole in the aortic root affecting aortic valve leaflet cell and function Panel **a**. Panel **b**, demonstrates the cellular architecture of a normal aortic valve. Panel **c**, demonstrates the osteogenic phenotype of the calcified aortic valve [1] (Permission obtained to reproduce the figure)

Cardiac Valve Cell Types: Valvular Interstitial Cells

VICs are abundant in all layers of the heart valves and are crucial to function. VICs synthesize extracellular matrix proteins, and express matrix degrading enzymes, which regulate remodeling of collagen and other matrix components [37]. They modulate function among phenotypes in response to changes in stimulation by the mechanical environment or by certain chemicals, during valvular homeostasis, adaptation, and pathology [38].

Valvular Endothelial Cells

VECs resemble endothelial cells elsewhere in the circulation in some respects. However, they are phenotypically different from vascular endothelial cells in the adjacent aorta and elsewhere in the circulation secondary to embryologic origin [39]. VECs likely interact with VICs to maintain the integrity of valve tissues and potentially mediate disease. Evidence indicates that different transcriptional profiles are expressed by VECs on the opposite (i.e., aortic and ventricular) faces of a normal adult pig aortic valve, and some investigators have hypothesized that these differences may contribute to the typical localization of early pathologic aortic valve calcification predominantly near the outflow surface secondary to inhibitors along the inflow surface [40]. Studies indicate that abnormal hemodynamic forces (such as hypertension [41], elevated stretch [42], or shear stresses [42]) experienced by the valve leaflets can cause tissue remodeling and inflammation, which may lead to calcification, stenosis, and ultimate valve failure.

Osteogenic Phenotype

Calcific aortic valve stenosis has characteristic pathological features [4]. The calcific process begins deep in the valvular tissue, near the margins of attachment. Over time nodules develop over the aortic surface of the valve leaflet. Lipids also play a role in the cell signaling of vascular calcification [43]. Studies in the field of vascular calcification have played an important role in recent experimental studies in valvular heart disease. Surgical pathological studies have shown the presence of oxidized LDL in calcified valves [12, 44]. Patients with homozygous familial hypercholesterolemia (FH) provide an opportunity to test the hypothesis that lipids play a role in the development of calcific aortic stenosis because these patients have extremely elevated levels of low-density lipoprotein cholesterol (LDL) without other traditional risk factors for coronary artery disease [45–48]. Angiotensin converting enzyme (ACE) is expressed and colocalizes with LDL in calcified aortic valves [49], and slows progression in a small observational study [50].

Initiating Events: Oxidative Stress

In the presence of cardiovascular risk factors, similar to vascular atherosclerosis, an early event is abnormalities in oxidative stress. This has been demonstrated in abnormal endothelial nitric oxide synthase function, which decreases normal physiologic levels of nitric oxide along the valve endothelium [51]. In atherosclerotic plaques, increased oxidative stress, causes an increase in NAD(P)H oxidase activity, similar to vascular atherosclerosis [52]. In calcified stenotic human [53] and mouse aortic valves [54], levels of superoxide and hydrogen peroxide are markedly increased. In addition, uncoupling of nitric oxide synthase [51] may play an important role in generation of superoxide in calcified aortic valves similar to the vasculature.

Calcifying Phenotype: Myofibroblast Differentiation to Bone

The initial confirmation of pathologic bone in the aortic valve was demonstrated by bone histomorphometry [5] and osteogenic gene expression [4] in diseased human valves. The likely sources of the myofibroblasts and osteoblasts that appear and persist in CAVD include native VICs, which contain mesenchymal progenitor-like cells that are highly plastic [55], and small numbers of circulating progenitors [56] and mesenchymal cells that transition from endothelial cells [57]. Statins represent a particularly intriguing avenue to pursue in regulating VIC function, as these drugs have demonstrated clinical but controversial slowing the progression of CAVD [58].

Cell-Cell Signaling in Calcific Aortic Valve Disease

The cell biology of the aortic valve is regulated by cell-cell communication between valve endothelial cells and valve interstitial cells (VICs). This specific cell communication, maintains the health of the valve and also is responsible for mediating valve disease. The "stem cell niche". provides a cellular architecture and also a gradient secondary to abnormal oxidative stress to initiate osteogenesis in the aortic valve [15] Potential triggers for VIC differentiation secondary to endothelial dysfunction include: hemodynamic shear stress, solid tissue stresses, reactive oxygen species, inflammatory cytokines and growth factors, and the cellular environment caused by other disease states, such as metabolic syndrome, diabetes mellitus, hypercholesterolemia, chronic renal disease, and disorders of calcium or phosphate metabolism. Once activated, VICs can differentiate into a variety of other cell types, including myofibroblasts and osteoblasts, although valve osteoblasts may respond to cellular signals differently than skeletal osteoblasts [15, 53, 54].

Garg et al. [32], discovered that a loss of function mutation in Notch1 was associated with accelerated aortic valve calcification and a number of congenital heart abnormalities. Normal Notch1 receptor regulates inhibition of osteoblastogenesis [59, 60]. The Notch1 splicing may be the regulatory switch important for the activation of the Wnt pathway and downstream calcification in these diseased valves [60–62]. The concept that cell-cell communication within a stem cell niche is necessary for the development of valvular heart disease, provides a foundation for the cell architecture, risk factors and the gradient involved during the initiation phase of oxidative stress in the aortic valve. The two corollaries necessary for an adult stem cell niche is to first define the physical architecture of the stem-cell niche and second is to define the gradient of proliferation to differentiation within the stem-cell niche. The endothelial lining cell located along the aortic surface is responsible for the secretion of a growth factors [63]. These cells interact with the subendothelial cells that are resident below the endothelial layer of cells. These cells have been characterized as myofibroblast cells [64–66].

In the aortic valve the communication for the stem cell niche is between the aortic valve endothelial cell and the adjacent myofibroblast cell located below the aortic lining endothelial cell as shown in Fig. 3.4. A Wnt3a which is secreted from the aortic valve endothelial cells binds to the Lrp5 receptor on the responding mesenchymal cell, the cardiac valve myofibroblast [15, 64, 67]. This system is appealing because the responding mesenchymal cell is isolated from the anatomic region adjacent and immediately below that of the endothelial cells producing the growth factor activity along the fibrosa surface. Similar to the vascular biology of the vasculature, which also has an endothelial cell lining which communicates with the

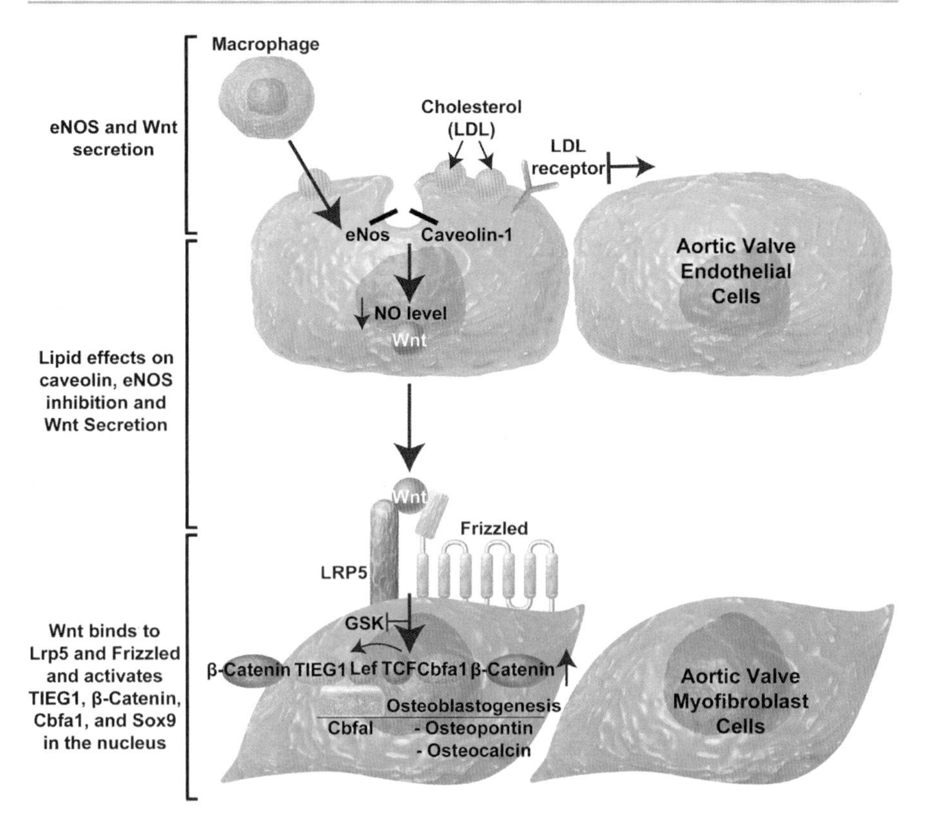

Fig. 3.4 Cell-Cell Signaling in CAVD. The top cell layer of the aortic valve is the endothelial layer of the aortic valve: Effects of Oxidative Stress induces the secretion of Wnt3a. The second cell layer along the Aortic Valve fibrosa surface is the valve myofibroblast where calcified nodules develop. In the presence of oxidative stress secondary to traditional cardiovascular risk factors, Secretion of Wnt3a from the endothelium binds to the Lrp5 receptor. Formation of the Lrp5/Wnt3a/Frizzled trimeric receptor complex along the surface of the myofibroblast extracellular membrane then activates Wnt Signaling. Once Beta-catenin translocates to the nucleus, then transcriptional activation of the transcription factors: Cbfa1/Sox9/TIEG1 in the nucleus to regulate osteogenesis in the valve myofibroblast

vascular smooth muscle cells, such as the release of nitric oxide in the activation of c-GMP in vascular smooth muscle relaxation as shown in Fig. 3.4. The second corollary for identifying a stem cell niche is to define the gradient responsible for the proliferation to differentiation process. The main postulate for this corollary stems from the risk factor hypothesis for the development of aortic valve disease. If traditional atherosclerotic risk factors are necessary for the initiation of disease, then these risk factors are responsible for the gradient necessary for the differentiation of myofibroblast cells to become an osteoblast calcifying phenotype [26, 61, 62, 65, 66, 68, 69]. If traditional risk factors are responsible for the development of valvular heart disease, then an oxidative stress mechanism is important for the development of a gradient in this niche.

The stem cell niche in combination with previously published data [61, 62, 69] indicates that the Wnt/Lrp5 pathway is implicated in the bone differentiation process and HMG CoA Reductase agents can slow the progression of this disease by inhibiting Lrp5 expression. Adult tissues stem cells are a population of functionally undifferentiated cells, capable of (1) homing (2) proliferation, (3) producing differentiated progeny, (4) self-renewing, (5) regeneration, and (6) reversibility in the use of these options. Within this definition, stem cells are defined by virtue of their functional potential and not by a specific observable characteristic. Lrp5 is important in normal valve development [70], in this stem cell niche, reactivation of latent Lrp5 expression [61, 71], regulates osteoblastogenesis in these mesenchymal cells. The two corollary requirements necessary for an adult stem cell niche is to first define the physical architecture of the stem-cell niche and second is to define the gradient of proliferation to differentiation within the stem-cell niche. The aortic valve endothelial cell communicates with the myofibroblast cell to activate the myofibroblast to differentiate to form an osteoblast-like phenotype [4]. This concept is similar to the endothelial/mesenchymal transition critical in normal valve development [72].

Mouse Models of CAVD

Mouse models of hypercholesterolemia demonstrate various features of human CAVD at the molecular and organ levels, and at least one develops stenosis [15, 54, 73, 74]. But hypercholesterolemia is only one of several conditions, including other risk factors for atherosclerosis, and specific genetic mutations [32]—that contributes to aortic stenosis, and may not be the most common. Therapies developed in high-cholesterol animal models [61, 73, 75, 76] may fail in human clinical trials [8], unless the therapies target final common pathways leading to CAVD, which remain to be elucidated. The implications are important for the design of future clinical trials. [77]

Elevated LDL and its oxidative modification represent one of the major factors of CAVD in the clinical settings. Therefore, addressing the mechanisms of CAVD in hypercholesterolemic animal models is a reasonable and essential approach. Development of CAVD has been shown in both apoE and LDL receptor deficient mice [54, 56, 78]. Aortic valves in hypercholesterolemic mice and rabbits [51, 61, 76], characterized by thickened leaflets with macrophage-rich subendothelial lesions in early stages and formation of calcific deposits on the aortic site of the valve in late stages, reproduce key pathologic features found in human valve disease. In addition, clinicopathological studies of stenotic aortic valves in humans identified lesions similar to those in inflamed atherosclerotic plaques [12, 44]. Cholesterol lowering in such models improves various features associated with atherogenesis and aortic valve disease [51, 61, 76] [79]. These animal models are extremely important and need to be characterized further in regards to CAVD. However, these models also have limitations in that no one model recapitulates the human disease process but each published model to date provides information which points towards the studies in the human disease process.

Clinical Trials: HMG CoA-Reductase Pathway

The first randomized prospective study testing the effects of HMG CoA reductase inhibitors in aortic valve disease was published in 2005 [8]. In this double-blind, placebo-controlled trial, patients with calcific aortic stenosis were randomly assigned to receive either 80 mg of atorvastatin daily or a matched placebo. Aortic-valve stenosis and calcification were assessed with the use of Doppler echocardiography and helical computed tomography, respectively. The SALTIRE investigators demonstrated a trend in slowing the progression of the aortic valve stenosis but not a statistically significant study for primary end-points. The SALTIRE investigators concluded that intensive lipid-lowering therapy does not halt the progression of calcific aortic stenosis or induce its regression [8], and the reason for this negative trial might be the timing of therapy [80]

In the RAAVE trial, Moura et al. [28], performed a prospective trial of AS with Rosuvastatin targeting serum LDL, slowed progression of echo hemodynamic measurements, and improved inflammatory biomarkers providing the first clinical evidence for targeted therapy using an HMG CoA reductase inhibitors in patients with asymptomatic moderate AS [28]. The next clinical trial, SEAS, examined intensive lipid lowering with Simvastatin and Ezetimibe in Aortic Stenosis [30]. This trial was a randomized, double-blind trial involving 1873 patients with mild-to-moderate, asymptomatic aortic stenosis. Again, the investigators concluded that the medication did not reduce the composite outcome of combined aortic-valve events in patient with aortic stenosis including echo progression and vascular end-points. Finally, the most recent trial, Astronomer, randomized patients to Rosuvastatin versus placebo in patients with moderate aortic valve disease and bicuspid aortic valve disease. This study also did not demonstrate slowing of progression of this disease [29]. These four clinical trials have different results, which may be due a number of reasons including differences in trial designs, differences in enrollment criteria, differences in statin medication, or timing of therapy [77]. The future of clinical valve trials may need further analysis of the trial design, the type of medications and the duration of the trials, but for now there is no indication for statin therapy in patients with valvular heart disease to slow progression of this disease. However, treatment of all cardiovascular patients with risk factors remains appropriate according to the guidelines as described by the American Heart Association and American College of Cardiology.

Prospective clinical studies of CAVD are hampered by the typically slow and variable progression of the disease. Patients who present with aortic stenosis are already in the later stages of the disease. Echocardiography is the standard for evaluating the severity of aortic stenosis and is a useful surrogate endpoint for clinical studies in the later stages. CT is a relatively high-resolution and high-sensitivity technique for evaluating aortic valve calcium and is a useful endpoint for clinical studies in the earlier stages. However, molecular imaging, with sub-millimeter resolution, may be able to identify and study the mechanisms of even earlier subclinical aortic valve calcification. Current information does not yet support a specific pharmacological target or design of a large CAVD treatment clinical trial. Recent studies showing lipid reduction to be ineffective may have been limited by the late stage of the disease or by an insensitive measure of effect. Whether patients at an earlier stage, e.g., with aortic sclerosis, or with specific known risk factors such as BAV, should be treated with lipid lowering therapy, angiotensin converting enzyme inhibitors, or novel pharmacological interventions—even if they don't meet the current criteria for therapy—remains an open question.

NHLBI Recommendations for CAVD

Based on this review of the current state of knowledge as summarized in this paper, the Working Group made the following recommendations for future research on CAVD [1].

1. Identify genetic, anatomic, and clinical risk factors for the distinct phases of initiation and progression of CAVD, to identify individuals at higher risk, to determine interactions between risk factors, and to determine whether the severity of aortic stenosis is a risk factor for surgical aortic valve replacement. These factors should encompass the unique contributions of atherosclerosis, metabolic syndrome, hypercholesterolemia, type II diabetes, and chronic kidney disease. New, larger epidemiological studies and existing epidemiological datasets in which CT scans, echocardiograms, or possibly magnetic resonance imaging scans have been obtained, could be used in this effort.
2. Develop high-resolution and high-sensitivity imaging modalities that can identify early and subclinical CAVD, including molecular imaging and other innovative imaging approaches. Continue research to define the state-of-the-art for detecting early calcification not identified by traditional echocardiographic imaging.
3. Understand the pathogenesis and pathophysiology of bicuspid aortic valve, especially to establish correlations between phenotype and genotype, and to clarify the key features of this disease process that potentiate calcification.
4. Understand the basic valve biology (e.g., early events, mechanisms and regulatory effects) of CAVD, including signaling pathways and the roles of valve interstitial and endothelial cells and the autocrine and paracrine signaling between them, the extracellular matrix and matrix stiffness, the role of age-related changes in both valve cells and extracellular matrix, the interacting mechanisms of cardiovascular calcification and physiologic bone mineralization, and micro-scale mechanotransduction and macro-scale hemodynamics.
5. Develop and validate suitable multi-scale in vitro, ex vivo, and animal models. Improved models are needed that realistically duplicate the conditions in which human CAVD develops. Metabolic studies are needed, from the cellular level through the patient level, to define those conditions.
6. Identify the relationship between calcification of the aortic valve and bone and the reciprocal regulation of these processes.
7. Encourage, promote, or establish tissue banks that make valve tissue from surgery, pathology, and autopsy unsuitable or unneeded for transplantation—with and without CAVD—available for research. Human valve cell lines should be derived including immortalized VICs.

8. Conduct clinical studies specific to CAVD to determine the feasibility of earlier pharmacological intervention and to determine the risk factors and optimal timing of surgical valve replacement.

Summary

In Summary, Calcific Aortic Valve Disease is the number one indication for cardiac valve replacement. The cellular mechanisms are complex and evolving rapidly. There are no medical therapies established to slow the progression of this disease, but continued research into the cellular mechanisms will provide the foundation for future clinical trials to slow progression and delay surgical valve replacement.

References

1. Rajamannan NM, Evans FJ, Aikawa E, et al. Calcific aortic valve disease: not simply a degenerative process: a review and agenda for research from the National Heart and Lung and Blood Institute Aortic Stenosis Working Group. Executive summary: Calcific aortic valve disease-2011 update. Circulation. 2011;124:1783–91.
2. Nishimura RA, Otto CM, Bonow RO, et al. 2014 AHA/ACC guideline for the management of patients with valvular heart disease: executive summary: a report of the American College of Cardiology/American Heart Association Task Force on Practice Guidelines. J Am Coll Cardiol. 2014;63:2438–88.
3. Rajamannan NM, Nealis TB, Subramaniam M, et al. Calcified rheumatic valve neoangiogenesis is associated with vascular endothelial growth factor expression and osteoblast-like bone formation. Circulation. 2005;111:3296–301.
4. Rajamannan NM, Subramaniam M, Rickard D, et al. Human aortic valve calcification is associated with an osteoblast phenotype. Circulation. 2003;107:2181–4.
5. Mohler ER 3rd, Gannon F, Reynolds C, Zimmerman R, Keane MG, Kaplan FS. Bone formation and inflammation in cardiac valves. Circulation. 2001;103:1522–8.
6. Stewart BF, Siscovick D, Lind BK, et al. Clinical factors associated with calcific aortic valve disease. Cardiovascular Health Study. J Am Coll Cardiol. 1997;29:630–4.
7. Figueiredo CP, Rajamannan NM, Lopes JB, et al. Serum phosphate and hip bone mineral density as additional factors for high vascular calcification scores in a community-dwelling: the Sao Paulo Ageing & Health Study (SPAH). Bone. 2013;52:354–9.
8. Cowell SJ, Newby DE, Prescott RJ, et al. A randomized trial of intensive lipid-lowering therapy in calcific aortic stenosis. N Engl J Med. 2005;352:2389–97.
9. Otto CM, Lind BK, Kitzman DW, Gersh BJ, Siscovick DS. Association of aortic-valve sclerosis with cardiovascular mortality and morbidity in the elderly. N Engl J Med. 1999;341:142–7.
10. Owens DS, Katz R, Takasu J, Kronmal R, Budoff MJ, O'Brien KD. Incidence and progression of aortic valve calcium in the Multi-ethnic Study of Atherosclerosis (MESA). Am J Cardiol. 2010;105:701–8.
11. Owens DS, Budoff MJ, Katz R, et al. Aortic valve calcium independently predicts coronary and cardiovascular events in a primary prevention population. JACC Cardiovasc Imaging. 2012;5:619–25.
12. O'Brien KD, Reichenbach DD, Marcovina SM, Kuusisto J, Alpers CE, Otto CM. Apolipoproteins B, (a), and E accumulate in the morphologically early lesion of 'degenerative' valvular aortic stenosis. Arterioscler Thromb Vasc Biol. 1996;16:523–32.
13. Otto CM, Kuusisto J, Reichenbach DD, Gown AM, O'Brien KD. Characterization of the early lesion of 'degenerative' valvular aortic stenosis. Histological and immunohistochemical studies. Circulation. 1994;90:844–53.

14. Rajamannan NM. Calcific aortic stenosis: lessons learned from experimental and clinical studies. Arterioscler Thromb Vasc Biol. 2009;29:162–8.
15. Rajamannan NM. Oxidative-mechanical stress signals stem cell niche mediated Lrp5 osteogenesis in eNOS(−/−) null mice. J Cell Biochem. 2012;113:1623–34.
16. Mohler ER 3rd, Adam LP, McClelland P, Graham L, Hathaway DR. Detection of osteopontin in calcified human aortic valves. Arterioscler Thromb Vasc Biol. 1997;17:547–52.
17. Thanassoulis G, Campbell CY, Owens DS, et al. Genetic associations with valvular calcification and aortic stenosis. N Engl J Med. 2013;368:503–12.
18. Thanassoulis G. Lipoprotein(a) in calcific aortic valve disease: from genomics to novel drug target for aortic stenosis. J Lipid Res. 2015. https://doi.org/10.1194/jlr.R051870.
19. Smith JG, Luk K, Schulz CA, et al. Association of low-density lipoprotein cholesterol-related genetic variants with aortic valve calcium and incident aortic stenosis. JAMA. 2014;312:1764–71.
20. Cao J, Steffen BT, Budoff M, et al. Lipoprotein(a) levels are associated with subclinical calcific aortic valve disease in White and Black individuals: the multi-ethnic study of atherosclerosis. Arterioscler Thromb Vasc Biol. 2016;36:1003–9.
21. Bild DE, Detrano R, Peterson D, et al. Ethnic differences in coronary calcification: the multi-ethnic study of atherosclerosis (MESA). Circulation. 2005;111:1313–20.
22. Huang CC, Lloyd-Jones DM, Guo X, et al. Gene expression variation between African Americans and whites is associated with coronary artery calcification: the multiethnic study of atherosclerosis. Physiol Genomics. 2011;43:836–43.
23. Elmariah S, Delaney JA, O'Brien KD, et al. Bisphosphonate use and prevalence of valvular and vascular calcification in women MESA (the multi-ethnic study of atherosclerosis). J Am Coll Cardiol. 2010;56:1752–9.
24. Nasir K, Katz R, Al-Mallah M, et al. Relationship of aortic valve calcification with coronary artery calcium severity: the multi-ethnic study of atherosclerosis (MESA). J Cardiovasc Comput Tomogr. 2010;4:41–6.
25. Ix JH, Shlipak MG, Katz R, et al. Kidney function and aortic valve and mitral annular calcification in the multi-ethnic study of atherosclerosis (MESA). Am J Kidney Dis. 2007;50:412–20.
26. Tintut Y, Alfonso Z, Saini T, et al. Multilineage potential of cells from the artery wall. Circulation. 2003;108:2505–10.
27. Rajamannan NM. Embryonic cell origin defines functional role of Lrp5. Atherosclerosis. 2014;236:196–7.
28. Moura LM, Ramos SF, Zamorano JL, et al. Rosuvastatin affecting aortic valve endothelium to slow the progression of aortic stenosis. J Am Coll Cardiol. 2007;49:554–61.
29. Chan KL, Teo K, Dumesnil JG, Ni A, Tam J, Investigators A. Effect of Lipid lowering with rosuvastatin on progression of aortic stenosis: results of the aortic stenosis progression observation: measuring effects of rosuvastatin (ASTRONOMER) trial. Circulation. 2010;121:306–14.
30. Rossebo AB, Pedersen TR, Boman K, et al. Intensive lipid lowering with simvastatin and ezetimibe in aortic stenosis. N Engl J Med. 2008;359:1343–56.
31. Capoulade R, Chan KL, Yeang C, et al. Oxidized phospholipids, lipoprotein(a), and progression of calcific aortic valve stenosis. J Am Coll Cardiol. 2015;66:1236–46.
32. Garg V, Muth AN, Ransom JF, et al. Mutations in NOTCH1 cause aortic valve disease. Nature. 2005;437:270–4.
33. Roberts WC, Ko JM. Frequency by decades of unicuspid, bicuspid, and tricuspid aortic valves in adults having isolated aortic valve replacement for aortic stenosis, with or without associated aortic regurgitation. Circulation. 2005;111:920–5.
34. Lee TC, Zhao YD, Courtman DW, Stewart DJ. Abnormal aortic valve development in mice lacking endothelial nitric oxide synthase. Circulation. 2000;101:2345–8.
35. Lazaros G, Toutouzas K, Drakopoulou M, Boudoulas H, Stefanadis C, Rajamannan N. Aortic sclerosis and mitral annulus calcification: a window to vascular atherosclerosis? Expert Rev Cardiovasc Ther. 2013;11:863–77.
36. Schoen FJ. Evolving concepts of cardiac valve dynamics: the continuum of development, functional structure, pathobiology, and tissue engineering. Circulation. 2008;118:1864–80.

37. Liu AC, Joag VR, Gotlieb AI. The emerging role of valve interstitial cell phenotypes in regulating heart valve pathobiology. Am J Pathol. 2007;171:1407–18.
38. Aikawa E, Whittaker P, Farber M, et al. Human semilunar cardiac valve remodeling by activated cells from fetus to adult: implications for postnatal adaptation, pathology, and tissue engineering. Circulation. 2006;113:1344–52.
39. Davies PF, Passerini AG, Simmons CA. Aortic valve: turning over a new leaf(let) in endothelial phenotypic heterogeneity. Arterioscler Thromb Vasc Biol. 2004;24:1331–3.
40. Simmons CA, Grant GR, Manduchi E, Davies PF. Spatial heterogeneity of endothelial phenotypes correlates with side-specific vulnerability to calcification in normal porcine aortic valves. Circ Res. 2005;96:792–9.
41. Xing Y, Warnock JN, He Z, Hilbert SL, Yoganathan AP. Cyclic pressure affects the biological properties of porcine aortic valve leaflets in a magnitude and frequency dependent manner. Ann Biomed Eng. 2004;32:1461–70.
42. Balachandran K, Sucosky P, Jo H, Yoganathan AP. Elevated cyclic stretch alters matrix remodeling in aortic valve cusps: implications for degenerative aortic valve disease. Am J Physiol. 2009;296:H756–64.
43. Demer LL. Cholesterol in vascular and valvular calcification. Circulation. 2001;104:1881–3.
44. Olsson M, Thyberg J, Nilsson J. Presence of oxidized low density lipoprotein in nonrheumatic stenotic aortic valves. Arterioscler Thromb Vasc Biol. 1999;19:1218–22.
45. Goldstein JL, Brown MS. Familial hypercholesterolemia: identification of a defect in the regulation of 3-hydroxy-3-methylglutaryl coenzyme A reductase activity associated with overproduction of cholesterol. Proc Natl Acad Sci U S A. 1973;70:2804–8.
46. Sprecher DL, Schaefer EJ, Kent KM, et al. Cardiovascular features of homozygous familial hypercholesterolemia: analysis of 16 patients. Am J Cardiol. 1984;54:20–30.
47. Kawaguchi A, Miyatake K, Yutani C, et al. Characteristic cardiovascular manifestation in homozygous and heterozygous familial hypercholesterolemia. Am Heart J. 1999;137:410–8.
48. Rajamannan NM, Edwards WD, Spelsberg TC. Hypercholesterolemic aortic-valve disease. N Engl J Med. 2003;349:717–8.
49. O'Brien KD, Shavelle DM, Caulfield MT, et al. Association of angiotensin-converting enzyme with low-density lipoprotein in aortic valvular lesions and in human plasma. Circulation. 2002;106:2224–30.
50. Shavelle DM, Takasu J, Budoff MJ, Mao S, Zhao XQ, O'Brien KD. HMG CoA reductase inhibitor (statin) and aortic valve calcium. Lancet. 2002;359:1125–6.
51. Rajamannan NM, Subramaniam M, Stock SR, et al. Atorvastatin inhibits calcification and enhances nitric oxide synthase production in the hypercholesterolaemic aortic valve. Heart. 2005;91:806–10.
52. Fukai T, Galis ZS, Meng XP, Parthasarathy S, Harrison DG. Vascular expression of extracellular superoxide dismutase in atherosclerosis. J Clin Invest. 1998;101:2101–11.
53. Miller JD, Chu Y, Brooks RM, Richenbacher WE, Pena-Silva R, Heistad DD. Dysregulation of antioxidant mechanisms contributes to increased oxidative stress in calcific aortic valvular stenosis in humans. J Am Coll Cardiol. 2008;52:843–50.
54. Weiss RM, Ohashi M, Miller JD, Young SG, Heistad DD. Calcific aortic valve stenosis in old hypercholesterolemic mice. Circulation. 2006;114:2065–9.
55. Chen JH, Yip CY, Sone ED, Simmons CA. Identification and characterization of aortic valve mesenchymal progenitor cells with robust osteogenic calcification potential. Am J Pathol. 2009;174:1109–19.
56. Tanaka K, Sata M, Fukuda D, et al. Age-associated aortic stenosis in apolipoprotein E-deficient mice. J Am Coll Cardiol. 2005;46:134–41.
57. Paranya G, Vineberg S, Dvorin E, et al. Aortic valve endothelial cells undergo transforming growth factor-beta-mediated and non-transforming growth factor-beta-mediated transdifferentiation in vitro. Am J Pathol. 2001;159:1335–43.
58. Antonini-Canterin F, Hirsu M, Popescu BA, et al. Stage-related effect of statin treatment on the progression of aortic valve sclerosis and stenosis. Am J Cardiol. 2008;102:738–42.

59. Sciaudone M, Gazzerro E, Priest L, Delany AM, Canalis E. Notch 1 impairs osteoblastic cell differentiation. Endocrinology. 2003;144:5631–9.
60. Deregowski V, Gazzerro E, Priest L, Rydziel S, Canalis E. Notch 1 overexpression inhibits osteoblastogenesis by suppressing Wnt/beta-catenin but not bone morphogenetic protein signaling. J Biol Chem. 2006;281:6203–10.
61. Rajamannan NM, Subramaniam M, Caira F, Stock SR, Spelsberg TC. Atorvastatin inhibits hypercholesterolemia-induced calcification in the aortic valves via the Lrp5 receptor pathway. Circulation. 2005;112:I229–34.
62. Shao JS, Cheng SL, Pingsterhaus JM, Charlton-Kachigian N, Loewy AP, Towler DA. Msx2 promotes cardiovascular calcification by activating paracrine Wnt signals. J Clin Invest. 2005;115:1210–20.
63. Rajamannan NM, Helgeson SC, Johnson CM. Anionic growth factor activity from cardiac valve endothelial cells: partial purification and characterization. Clin Res 1988:309A.
64. Johnson CM, Hanson MN, Helgeson SC. Porcine cardiac valvular subendothelial cells in culture: cell isolation and growth characteristics. J Mol Cell Cardiol. 1987;19:1185–93.
65. Mohler ER 3rd, Chawla MK, Chang AW, et al. Identification and characterization of calcifying valve cells from human and canine aortic valves. J Heart Valve Dis. 1999;8:254–60.
66. Osman L, Yacoub MH, Latif N, Amrani M, Chester AH. Role of human valve interstitial cells in valve calcification and their response to atorvastatin. Circulation. 2006;114:I547–52.
67. Johnson CM, Helgeson SC. Glycoproteins synthesized by cultured cardiac valve endothelial cells: unique absence of fibronectin production. Biochem Biophys Res Commun. 1988;153:46–50.
68. Wada T, McKee MD, Steitz S, Giachelli CM. Calcification of vascular smooth muscle cell cultures: inhibition by osteopontin. Circ Res. 1999;84:166–78.
69. Kirton JP, Crofts NJ, George SJ, Brennan K, Canfield AE. Wnt/beta-catenin signaling stimulates chondrogenic and inhibits adipogenic differentiation of pericytes: potential relevance to vascular disease? Circ Res. 2007;101:581–9.
70. Hurlstone AF, Haramis AP, Wienholds E, et al. The Wnt/beta-catenin pathway regulates cardiac valve formation. Nature. 2003;425:633–7.
71. Caira FC, Stock SR, Gleason TG, et al. Human degenerative valve disease is associated with up-regulation of low-density lipoprotein receptor-related protein 5 receptor-mediated bone formation. J Am Coll Cardiol. 2006;47:1707–12.
72. Paruchuri S, Yang JH, Aikawa E, et al. Human pulmonary valve progenitor cells exhibit endothelial/mesenchymal plasticity in response to vascular endothelial growth factor-A and transforming growth factor-beta2. Circ Res. 2006;99:861–9.
73. Rajamannan NM. Atorvastatin attenuates bone loss and aortic valve atheroma in LDLR mice. Cardiology. 2015;132:11–5.
74. Rajamannan NM. The role of Lrp5/6 in cardiac valve disease: experimental hypercholesterolemia in the ApoE−/− /Lrp5−/− mice. J Cell Biochem. 2011;112:2987–91.
75. Makkena B, Salti H, Subramaniam M, et al. Atorvastatin decreases cellular proliferation and bone matrix expression in the hypercholesterolemic mitral valve. J Am Coll Cardiol. 2005;45:631–3.
76. Rajamannan NM, Subramaniam M, Springett M, et al. Atorvastatin inhibits hypercholesterolemia-induced cellular proliferation and bone matrix production in the rabbit aortic valve. Circulation. 2002;105:2260–5.
77. Rajamannan NM. Mechanisms of aortic valve calcification: the LDL-density-radius theory: a translation from cell signaling to physiology. Am J Physiol. 2010;298:H5–15.
78. Aikawa E, Nahrendorf M, Sosnovik D, et al. Multimodality molecular imaging identifies proteolytic and osteogenic activities in early aortic valve disease. Circulation. 2007;115:377–86.
79. Miller JD, Weiss RM, Serrano KM, et al. Lowering plasma cholesterol levels halts progression of aortic valve disease in mice. Circulation. 2009;119:2693–701.
80. Newby DE, Cowell SJ, Boon NA. Emerging medical treatments for aortic stenosis: statins, angiotensin converting enzyme inhibitors, or both? Heart. 2006;92:729–34.
81. Rajamannan NM, Moura L. The lipid hypothesis in calcific aortic valve disease: the role of the multi-ethnic study of atherosclerosis. Arterioscler Thromb Vasc Biol. 2016;36:774–6.

Osteocardiology: Calcific Aortic Disease

4

Introduction

Diseases of the aorta include aneurysms, dissection, and atheroma calcification. The incidence is increasing with the aging global population and the increase sensitivity of diagnostic imaging. Understanding the risk factors associated with the development of calcific aortic disease, will help to identify future patients at risk and also slow progression of this disease before it is too late. Atherosclerosis plays a major role in the diseases of the thoracic ascending and descending, and the abdominal aorta. Atherosclerosis can result in weakening of the abdominal wall making it prone to aneurysm formation, dissection or chronic calcification in the aortic wall. The development of aortic atherosclerosis is associated with traditional cardiovascular risk factors including smoking, hypertension, diabetes, and elevated cholesterol levels. Atherosclerosis of the aorta can lead to formation of complex atheroma plaques, which can result in embolization causing cerebral and peripheral artery occlusive events. Other causes of aortic diseases include inflammatory, genetic, trauma and dissection, are important causes of aortic diseases. Similar to coronary artery calcification and calcific aortic valve disease, this chapter will focus on atherosclerotic calcification.

MESA and Calcific Aortic Disease (CAD)

In 2008, Investigators measured the calcific aortic disease (CAD) including ascending and descending thoracic aortic calcification in the MESA cohort. To determine which of cardiovascular risk and ethnicity variables are independently associated with thoracic calcium [1].

The Multi-Ethnic Study of Atherosclerosis (MESA) study population included a population based sample of four ethnic groups (12% Chinese, 38% White, 22% Hispanic and 28% black) of 6814 women and men ages 45–84 years old. In this

N.M. Rajamannan, *Osteocardiology*, DOI 10.1007/978-3-319-64994-8_4

study, the investigators quantified CAD, which ranged from the lower edge of the pulmonary artery bifurcation to the cardiac apex.

The overall prevalence of CAD was 28.0%. In the ethnic groups, prevalence of CAD was 32.4% Chinese, 32.4% White, 24.9% Hispanic and 22.4% Black. All age categories of females had a higher prevalence of thoracic calcification than males (total age prevalence: 29.1% and 26.8%, respectively). CAD were most strongly associated with hypertension and current smoking. In addition, diabetes, hypercholesterolemia, high LDL, low HDL, family history of heart attack and high CRP were all associated with increased CAD. Overall p-value for difference between genders for prevalence of CAD = 0.037. Overall p-value for difference between race for prevalence of CAD <0.001. The only significant gender differences distributed by race were for Chinese (p = 0.035) and Hispanic (p = 0.042) participants. The investigators concluded that factors for aortic calcification were similar to cardiovascular risk factors in a large population based cohort [1]. This study is the first to determine parallel risk factors for traditional cardiovascular risk factors and risk factors for CAD, with a long term subclinical phase until it reaches clinical overt disease as shown in Fig. 4.1.

In 2010, the Investigators took the study one step further, and compared the calcification in the thoracic aorta to the coronary arteries in the MESA cohort [2]. The study indicates that CAD is a significant predictor of future coronary events only in women, independent of coronary artery calcification (CAC). The mean age of the study population (n = 6807) was 62 ± 10 years (47% males). At baseline, the prevalence of CAD and CAC was 28% (1904/6809) and 50% (3393/6809), respectively. Over 4.5 ± 0.9 years, a total of 232 participants (3.41%) had CHD events, of which 132 (1.94%) had a hard event (myocardial infarction, resuscitated cardiac arrest, or CHD death). There was a significant interaction between gender and CAD for cardiovascular heart disease CHD events (p < 0.05). Specifically, in women, the risk of all CHD event was nearly threefold greater among those with CAD (hazard ratio: 3.04, 95% CI; 1.60–5.76).

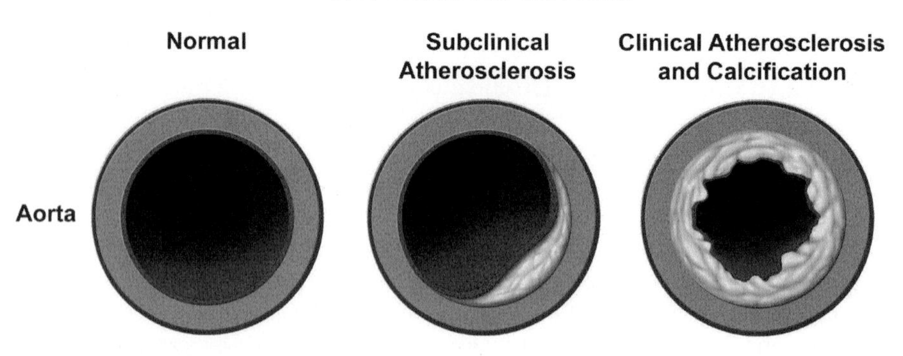

Fig. 4.1 Progression of calcific aortic disease

MESA Defines Calcific Aortic Disease and Bone Disease

Over the past 20 years, studies have shown that there are parallel risk factors associated with atherosclerosis and osteoporosis, including gender, lipids, hypertension, smoking etc. The MESA study, in 2008 [3], confirmed the independent association between volumetric trabecular bone mineral density (vBMD) of the lumbar spine and coronary artery calcification (CAC) and calcific aortic disease (CAD). Women had a prevalence of CAC of 47% as compared to men who had 68% coronary calcification. There was lower vBMD and greater CAC in women ($p < 0.002$) and greater CAD among women ($p = 0.004$) and men ($p < 0.001$). After adjustment, vBMD was inversely associated with CAC prevalence in women and CAD prevalence in men. The authors concluded that the results were modest, but significant to demonstrate an independent association between atherosclerosis and bone loss and that the two may be related.

Familial Hypercholesterolemia and Calcific Aortic Disease

Familial hypercholesterolemia (FH) is the most common monogenic disorder associated with premature atherosclerotic cardiovascular disease, calcific aortic valve disease, and cerebrovascular disease. Aortic calcification is a long-term complication secondary to FH. There are three types of calcification process which have been described (1), atherosclerosis associated intimal calcification of the intima (2), medial calcification/Monckeberg type of sclerosis, and (3) genetic disorder related calcification [4]. For centuries, pathologists have described the presence of bone in the histopathologic description of calcified arteries and valves. However, it is not until the last 20 years that the risk factors, and cellular mechanisms have revealed that this bone formation process is an active biology and not passive degeneration.

Defining Aorta Calcification in Familial Hypercholesterolemia

McGill University has defined calcification in the aorta in their cohort of patients with the diagnosis of Familial Hypercholesterolemia. They measure the degree of aortic calcification in heterozygous FH (heterozygote FH) compared to both (homozygote FH) and controls. They demonstrated that LDLR gene contributes to aortic calcifications in a gene-dosage effect [5].

In a recent study by Kindi et al. [6], the investigators sought to determine the rate of progression of aorta calcification in patients with HeFH. Sixteen HeFH patients, all with the null LDLR DEL15Kb mutation were studied using thoraco-abdominal CT scans and quantification scores. Patients were scanned at baseline and rescanned an average of 8.2 ± 0.8 years after the first scan. Mean LDL-C was 2.53 mmol/L on medical therapies. Aortic calcification increased in all patients in an exponential fashion with respect to age. Age was the strongest correlate of AoCa score. Investigators studied only patients heterozygous for FH and analyzed the data using age statistical analysis scores. The effect of age demonstrated a fivefold increase in the progression in the HeFH patient population with an exponential increase only correlating with age as a risk factor. Despite the known defect in lipid metabolism in this subset of FH patients, traditional risk factor of lipids did not play a role in the progression of the calcification process in this patient population. Furthermore, the effect of lipid lowering of 65% reduction in the LDL-C from the baseline values, medical therapy did not attenuate the process. The results of this study indicate that patients who have the diagnosis of HeFH, and are delayed to the time of diagnosis, the calcification process will progress even on optimal medical therapy to lower lipids.

To further understand this important scientific finding, the role of cellular mechanism of calcification can begin to help to determine future approaches for this patient population. Calcification in the heart has been described as an osteogenic bone formation process [4, 7]. The discovery of the Lrp5 receptor in the gain of function [8] and loss of function [9] mutations in bone diseases, resulted in a number of studies which have shown that activation of the canonical Wnt pathway is important in osteoblastogenesis [10–13]. Studies in the field of cardiovascular medicine have also demonstrated that Lrp5 pathway is active in the calcification of arteries [14] and valves [15], and that the LDLR null mouse model expression of Lrp5 in calcifying valves [16] and arteries [14], which translates to the bone and lipid biology in patients with Familial Hypercholesterolemia [17, 18].

In Fig. 4.2, demonstrates calcification in the valve and in the aorta in patients with HeFH, Panel a, Control, Panel b HeFH. The mechanism evaluated in the LDLR

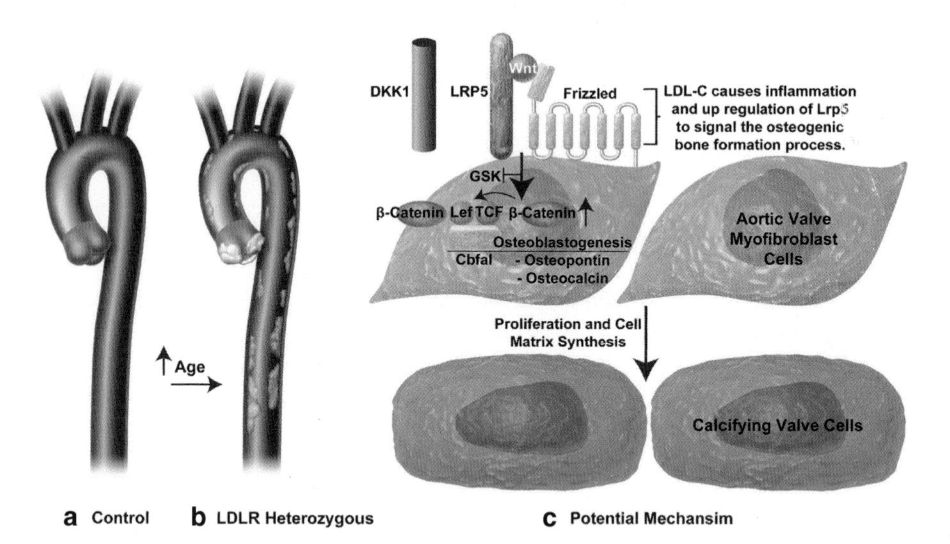

Fig. 4.2 Signaling mechanism in calcific aortic disease [24]. (Permission obtained to reproduce Figure). Panel **a**: Control Aorta in patients with normal cholesterol. Panel **b**: Calcified aorta and aortic valve in heterozygous FH patients. Panel **c**: Potential cellular mechanism of Lrp5/Wnt mediated bone formation in the valve and in the aorta

null mouse model, implicating the role of Lrp5 bone formation in the valve and in the aorta as shown in Panel c. The role of lipids in this disease process is critical in the initiating events for the disease process. Initiation of early atherosclerosis involves abnormal oxidative stress and release of Wnt [19]. Wnt then binds to Lrp5 to activate differentiation of the valve and vascular interstitial cells to become osteogenic bone forming cells [19], which over time calcifies and forms bone. In this study, the investigators have identified patients with HeFH as early as possible in the disease process and obtained baseline lipid levels, scans and initiated aggressive medical therapy to optimized lipid levels. However, identifying heterozygous patients with FH is not as easy as most patients are asymptomatic early in life. The age range at presentation indicates that exposure to high lipids over the lifetime of this patient population may be critical in the early initiation of the bone formation process, and once the process has occurred the effects of lipids and or statin therapies on the already formed bone have proven non consequential, or the point of no return in terms of modifying the disease process [20, 21]. The current treatment strategy for severe aortic calcification is surgery, which is a difficult surgical procedure especially when patients have developed porcelain aortic calcification [22].

Summary

The current Canadian Cardiovascular Society Guidelines [23] have recommended aggressive screening for patients with HeFH, including (1) increasing awareness of FH among health care providers and patients; (2) creating a national registry for FH individuals; (3) setting standards for screening, including cascade screening in affected families; (4) ensuring availability of standard-of-care therapies, in particular optimization of plasma low-density lipoprotein cholesterol levels and timely access to future validated therapies; (5) promoting patient-based support and advocacy groups; and (6) forming alliances with international colleagues, resources, and initiatives that focus on FH. These guidelines will set an aggressive foundation for the future clinical approach for Familial Hypercholesterolemia, and will help to act early to diagnose, treat, and slow progression of the complications of chronic hyperlipidemia, including calcific aortic disease.

References

1. Takasu J, Katz R, Nasir K, et al. Relationships of thoracic aortic wall calcification to cardiovascular risk factors: the multi-ethnic study of atherosclerosis (MESA). Am Heart J. 2008;155:765–71.
2. Budoff MJ, Nasir K, Katz R, et al. Thoracic aortic calcification and coronary heart disease events: the multi-ethnic study of atherosclerosis (MESA). Atherosclerosis. 2011;215:196–202.
3. Hyder JA, Allison MA, Wong N, et al. Association of coronary artery and aortic calcium with lumbar bone density: the MESA Abdominal Aortic Calcium Study. Am J Epidemiol. 2009;169:186–94.

4. Demer LL, Tintut Y. Inflammatory, metabolic, and genetic mechanisms of vascular calcification. Arterioscler Thromb Vasc Biol. 2014;34:715–23.
5. Alrasadi K, Alwaili K, Awan Z, Valenti D, Couture P, Genest J. Aortic calcifications in familial hypercholesterolemia: potential role of the low-density lipoprotein receptor gene. Am Heart J. 2009;157:170–6.
6. Kindi MA, Belanger AM, Sayegh K, et al. Aortic calcification progression in heterozygote familial hypercholesterolemia. Can J Cardiol. 2017;33:658–65.
7. Rajamannan NM, Subramaniam M, Rickard D, et al. Human aortic valve calcification is associated with an osteoblast phenotype. Circulation. 2003;107:2181–4.
8. Boyden LM, Mao J, Belsky J, et al. High bone density due to a mutation in LDL-receptor-related protein 5. N Engl J Med. 2002;346:1513–21.
9. Gong Y, Slee RB, Fukai N, et al. LDL receptor-related protein 5 (LRP5) affects bone accrual and eye development. Cell. 2001;107:513–23.
10. Fujino T, Asaba H, Kang MJ, et al. Low-density lipoprotein receptor-related protein 5 (LRP5) is essential for normal cholesterol metabolism and glucose-induced insulin secretion. Proc Natl Acad Sci U S A. 2003;100:229–34.
11. Babij P, Zhao W, Small C, et al. High bone mass in mice expressing a mutant LRP5 gene. J Bone Miner Res. 2003;18:960–74.
12. Westendorf JJ, Kahler RA, Schroeder TM. Wnt signaling in osteoblasts and bone diseases. Gene. 2004;341:19–39.
13. Holmen SL, Giambernardi TA, Zylstra CR, et al. Decreased BMD and limb deformities in mice carrying mutations in both Lrp5 and Lrp6. J Bone Miner Res. 2004;19:2033–40.
14. Awan Z, Denis M, Bailey D, et al. The LDLR deficient mouse as a model for aortic calcification and quantification by micro-computed tomography. Atherosclerosis. 2011;219:455–62.
15. Rajamannan NM. The role of Lrp5/6 in cardiac valve disease: experimental hypercholesterolemia in the ApoE−/− /Lrp5−/− mice. J Cell Biochem. 2011;112:2987–91.
16. Rajamannan NM. Atorvastatin attenuates bone loss and aortic valve atheroma in LDLR mice. Cardiology. 2015;132:11–5.
17. Rajamannan NM. Calcific aortic valve disease in familial hypercholesterolemia: the LDL-density-gene effect. J Am Coll Cardiol. 2015;66:2696–8.
18. ten Kate GJ, Bos S, Dedic A, et al. Increased aortic valve calcification in familial hypercholesterolemia: prevalence, extent, and associated risk factors. J Am Coll Cardiol. 2015;66:2687–95.
19. Rajamannan NM. Oxidative-mechanical stress signals stem cell niche mediated Lrp5 osteogenesis in eNOS(−/−) null mice. J Cell Biochem. 2012;113:1623–34.
20. Chan KL, Teo K, Dumesnil JG, Ni A, Tam J, Investigators A. Effect of Lipid lowering with rosuvastatin on progression of aortic stenosis: results of the aortic stenosis progression observation: measuring effects of rosuvastatin (ASTRONOMER) trial. Circulation. 2010;121:306–14.
21. Cowell SJ, Newby DE, Prescott RJ, et al. A randomized trial of intensive lipid-lowering therapy in calcific aortic stenosis. N Engl J Med. 2005;352:2389–97.
22. Grenon SM, Lachapelle K, Marcil M, Omeroglu A, Genest J, de Varennes B. Surgical strategies for severe calcification of the aorta (porcelain aorta) in two patients with homozygous familial hypercholesterolemia. Can J Cardiol. 2007;23:1159–61.
23. Primary P, Genest J, Hegele RA, et al. Canadian Cardiovascular Society position statement on familial hypercholesterolemia. Can J Cardiol. 2014;30:1471–81.
24. Rajamannan NM, Nattel S. Aortic vascular calcification: cholesterol lowering does not reduce progression in patients with familial hypercholesterolemia- or does it? Can J Cardiol. 2017;33:594–6.

Osteocardiology: Endochondral Bone Formation

5

Introduction

Bone the major component of the skeleton, is formed by two distinct ossification processes, intramembranous and endochondral. Intramembranous bone develops directly from mesenchymal cells condensing at ossification centers and differentiating directly into an osteoblast cell. The osteoblast cell is defined as a cell, which secretes matrix specific for bone formation. This ossification process gives rise to the flat bones of the skull, parts of the clavicle, and the periosteal surface of long bones. Endochondral ossification differs from the intramembranous component in that it is formed in the presence of a cartilaginous blastema. The cartilage blastema is well known as the formation critical for the development of mature cartilage cells.

The formation of mature cartilage is a complex multistep process requiring the sequential formation and degradation of cartilaginous structures, which serve as templates for the developing axial skeleton and appendicular bones. This formation of calcified bone on a cartilage scaffold occurs not only during skeletogenesis, but is an integral part of postnatal grown and fracture repair. In the early onset of skeletal development, undifferentiated mesenchymal cells come together to form condensations which have the shape of the skeletal elements that they will eventual configure. The next stage is followed by specific differential along either the osteoblastic (intramembranous) or the chondrocytic (endochondral pathway), See Fig. 5.1. Chondrocytes deposit specific extracellular matrix composes of various collagens, such as Type IIb, IX, and XI, which are cartilage-specific. Osteoblasts secrete proteins, which are osteoblast specific such as osteopontin, osteonectin, Type I collagen, with the cartilage matrix used as a scaffold for the bone formation.

N.M. Rajamannan, *Osteocardiology*, DOI 10.1007/978-3-319-64994-8_5

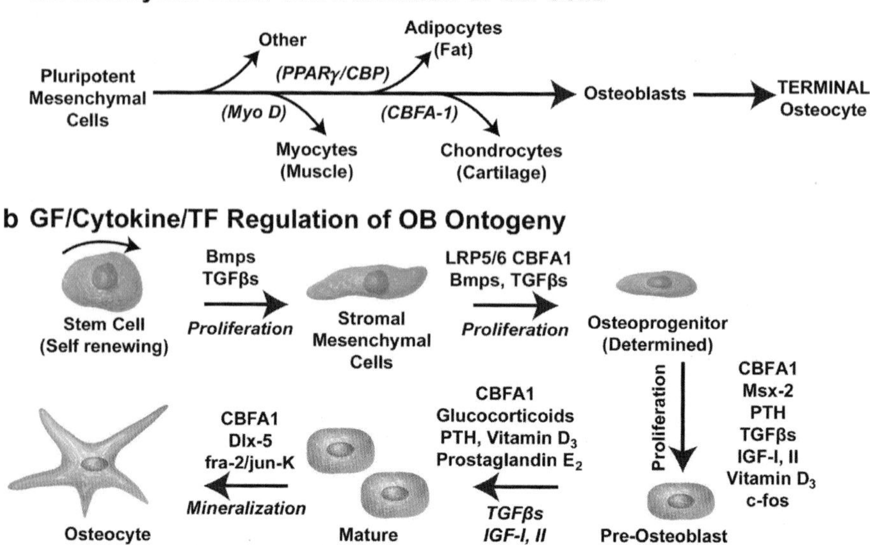

Fig. 5.1 Osteogenic Cascade demonstrates the differentiation pathway of mesenchymal cells to form bone

The Molecular Regulation of Mesenchymal Chondrogenesis to Osteogenesis

The Stages of Bone Formation

The initial stage is cell proliferation, as these cells exit the cell cycle and undergo further differentiation to a hypertrophic form, characterized by decreased expression of the cartilage proteins and increased expression of the calcification proteins, and apoptosis of the cells. Formation of the mineralized cartilage is vital as it favors the vascular invasion of the previously avascular cartilaginous condensations from the perichondrium. Osteoblasts, which originate from mesenchymal precursors, and osteoclasts, which are derived from the hematopoietic compartment, also enter the zone of cartilaginous hypertrophy, along with neoangiogenesis. The osteoclasts proceed to degrade the calcified cartilage matrix, while osteoblasts begin depositing the bone matrix, which consists of type I collagen, with the cartilage matrix being used as a scaffold, See Fig. 5.2.

The Osteoblast Cell

In the past, research surrounding the mechanism for bone formation focused on skeletal patterning. Recently, the focus of the signaling pathway has been on

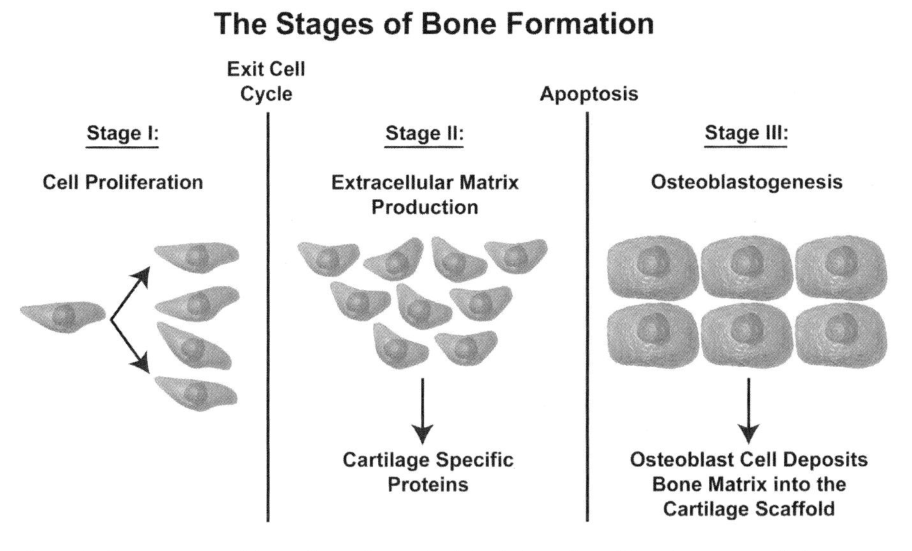

Fig. 5.2 Three stages of bone formation, which include cell proliferation, extracellular matrix production and osteogenic calcification

osteoblast cell biology and gene expression. Differentiation of the mesenchymal progenitor cells, chondrocytes and osteoblasts are critical towards understanding bone formation. Early progenitors are the skeletal cells derived from three distinct embryonic cell lineages: neural crest cells contribute to the craniofacial skeleton; sclerotome cells from somites give rise to the axial skeleton; and lateral plate mesoderm cells form the appendicular skeleton. Osteoblasts originate from immature mesenchymal cells, which could also give rise to chondrocytes, muscle, fat, ligament, and tendon cells.

These mesenchymal cells need to undergo transitional steps to becoming mature osteoblast cells. Figure 5.2, demonstrates a simplified progression to osteoblast differentiation. The transition, requires the activation or suppression of critical molecular elements for the progression of differentiation to occur. The key molecular switch for the induction of osteoblast formation is the activation of Cbfa1, (*core binding factor 1; Runx* 2). Cbfa1 is a transcription factors specific for osteogenesis. The expression and function is tightly controlled since appropriate activation and repression of its transcription during osteoblast differentiation would be essential for the regulation of the osteogenic gene cascade. Regulation of this master switch Cbfa1, have been reported to be due to three genes critical in the regulation of Cbfa1 expression: Msx2, when inactivated in mice leads to a down-regulation of Cbfa1; Bapx, a gene encoding a homeobox protein required for axial skeleton formation, which may activate Cbfa1, and Hoxa-2, another homeobox protein which inhibits Cbfa1 expression in the second branchial arch. Once activated these transcription factors are responsible for extracellular matrix production for the transition from Stage I to Stage II of the endochondral bone process.

During development of the long bone, growth plates localize either end of the skeletal element and the region of the cartilage is surrounded by perichondrium, which is composed of undifferentiated mesenchymal cells. The chondrocytes undergo several stages of differentiation. One transition from cell proliferation Stage I to hypertrophy, an event that precedes stage III. During chondrocyte hypertrophy us characterized by profound physical and biochemical changes, secondary to production of cartilage specific proteins critical for the enlargement of these cells.

Regulation of endochondral ossification occurs as the transition from Stage II to Stage III, critical in the osteogenesis phase of bone formation. Matrix vesicles are the initial sites of mineralization in the hypertrophic region of the growth plate and are critical components of the calcification process. The calcified matrix subsequently serves as a template for primary bone formation. Primary bone formation is initiated at the center of the cartilage template and results in the subsequent formation of two separate regions of endochondral bone, which develop at either end of the long bone. The growth plate is responsible for longitudinal growth of bones. Both chondrocyte proliferation and hypertrophy contribute to the lengthening of the limb. Because terminally differentiated hypertrophic cartilage is continuously replaced by bone, the tight regulation of the various steps of chondrocyte differentiation, particularly proliferation and hypertrophy, is critical for balancing the growth and ossification of the skeletal elements.

The Molecular Basis of Bone Formation in the Heart

Calcification is largely responsible for hemodynamic progression of aortic valve stenosis. Recent descriptive studies from patient specimens have demonstrated the cell changes associated with aortic valve calcification, including osteoblast expression, cell proliferation, and atherosclerosis [1–4]. Furthermore, these studies have also shown that specific bone cell phenotypes present in calcifying valve tissue from human specimens [5–9] demonstrate the potential for vascular cells to differentiate into calcifying phenotypes.

Recent observations in *ex vivo* human tissue suggest that rapid advancement in our understanding of the basic mechanisms involved in the initiation and progression of vascular and valvular calcification is now possible. If an osteoblast phenotype is present, then the factors important in the regulation of bone development and regeneration must be considered in the understanding of calcification of the aortic valve. It is well known that cardiovascular calcification is composed of hydroxyapatite deposited on a bone-like matrix of collagen, osteopontin (OP), and other minor bone matrix proteins [2, 10, 11], and regulation occurs via activation of specific transcription factors including MSX2 [12], Runx2 [5], and Sox9 [5]. Calcified aortic valves removed from surgical valve replacement show bone formation (osseous metaplasia) [2, 3, 5]. Further characterization of this phenotype has proven, in calcified bicuspid aortic valves, that immunohistochemistry staining shows the expression of osteopontin [10]. In addition, osteopontin expression has been demonstrated in the mineralization zones of heavily calcified aortic valves obtained at autopsy and surgery [1–3, 5].

Aortic Valve Bone Phenotype

The osteogenic phenotype in the aortic valve has been confirmed utilizing multiple modalities [3]. Contact microradiography and micro-computerized tomography were used to assess the 2-dimensional and 3-dimensional extent of mineralization. Mineralization borders were identified with von Kossa and Goldner's stains. Electron microscopy and energy-dispersive spectroscopy were performed for identification of bone ultrastructure and $CaPO_4$ composition. To analyze for the osteoblast and bone markers, reverse transcriptase-polymerase chain reaction was performed on calcified versus normal human valves for osteopontin, bone sialoprotein, osteocalcin, alkaline phosphatase, and the osteoblast-specific transcription factor Cbfa1. Microradiography and micro-computerized tomography confirmed the presence of calcification in the valve. Special stains for hydroxyapatite and $CaPO_4$ were positive in calcification margins. Electron microscopy identified mineralization, whereas energy-dispersive spectroscopy confirmed the presence of elemental $CaPO_4$. Reverse transcriptase-polymerase chain reaction revealed increased mRNA levels of osteopontin, bone sialoprotein, osteocalcin, and Cbfa1 in the calcified valves. This study is the first to determine the osteogenic gene expression profile in calcifying aortic valves, which confirmed the role of bone transcription factors regulating the disease mechanism [3]. Mohler et al., defined heterotopic ossification consisting of mature lamellar bone formation and active bone remodeling is a relatively common and unexpected finding in end-stage valvular heart disease and may be associated with repair of pathological microfractures in calcified cardiac valves [2]. Figure 5.3, shows the calcification in the aortic valve and mitral annulus.

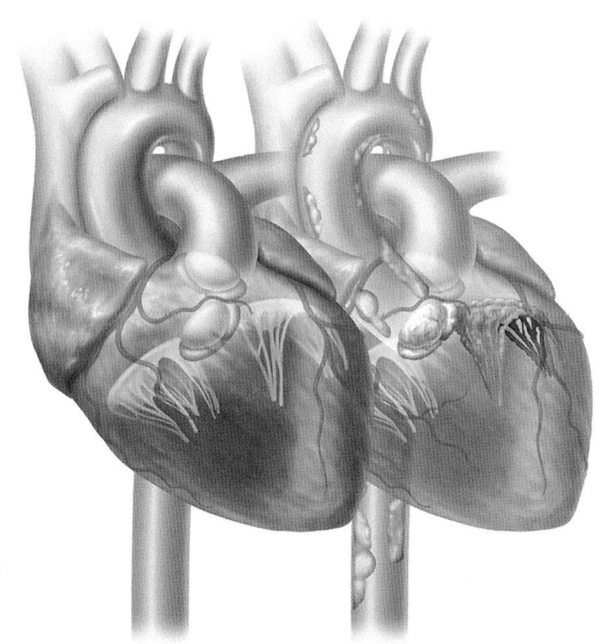

Fig. 5.3 Osteogenic phenotype demonstrating calcification in the valve, coronary artery and aorta

Coronary Artery Bone Phenotype

Early atherosclerosis begins with abnormal oxidative stress. Studies in vitro and in vivo have demonstrated that the role of oxidative stress is critical towards developing bone in the heart and osteoporosis in the femur. Studies have determined the role of lipoproteins in both processes, by examining the effect of minimally oxidized low-density lipoprotein, and several other lipid oxidation products on calcifying vascular cells and bone-derived preosteoblasts [9]. The investigators demonstrated that calcifying vascular cells developed cell proliferation and increase in bone matrix synthesis, a marker of osteoblast differentiation in the presence of oxidized lipoproteins. However, the oxidized lipoproteins inhibited the osteogenesis in the bone cells, confirming in vitro the hypothesis of the bone paradox [9, 13]. Studies determined whether these cells modulate vascular calcification in vitro, calcifying vascular cells (CVCs), a subpopulation of osteoblast-like cells derived from the artery wall, were cocultured with human peripheral blood monocytes for 5 days [13]. Results showed that alkaline phosphatase (ALP) activity, a marker of osteoblastic differentiation, was significantly greater in cocultures than in cultures of CVCs or monocytes alone. Both ALP activity and matrix mineralization increased in proportion to the number of monocytes added. Activation of monocyte/macrophages (M/Ms) by oxidized LDL further increased ALP activity in co-cultures, confirming the hypothesis in vitro of the bone paradox. In CVCs, MM-LDL but not native LDL inhibited proliferation, caused a dose-dependent increase in alkaline phosphatase activity, which is a marker of osteoblastic differentiation, and induced the formation of extensive areas of calcification. Similar to MM-LDL, oxidized 1-palmitoyl-2-arachidonoyl-sn-glycero-3-phosphorylcholine (ox-PAPC) and the isoprostane 8-iso prostaglandin E2 but not PAPC or isoprostane 8-iso prostaglandin F2 alpha induced alkaline phosphatase activity and differentiation of CVCs. In contrast, MM-LDL and the above oxidized lipids inhibited differentiation of the MC3T3-E1 bone cells, as evidenced by their stimulatory effect on proliferation and their inhibitory effect on the induction of alkaline phosphatase and calcium uptake. Data suggest that specific oxidized lipids may be the common factors underlying the pathogenesis of both atherosclerotic calcification and osteoporosis [9, 14].

Figure 5.3, shows the calcification in the coronary artery.

Calcific Aortic Bone Phenotype

To further understand this important scientific finding, the role of cellular mechanism of calcification can begin to help to determine future approaches for this patient population. Calcification in the heart has been described as an osteogenic bone formation process [3, 15]. The discovery of the Lrp5 receptor in the gain of function [16] and loss of function [17] mutations in bone diseases, resulted in a number of studies which have shown that activation of the canonical Wnt pathway is important in osteoblastogenesis [18–21]. Studies in the field of cardiovascular medicine have also demonstrated that Lrp5 pathway is active in the calcification of arteries [22] and valves [23], and that the LDLR null mouse model expression of Lrp5 in calcifying valves [24] and arteries [22], which translates to the bone and lipid biology in patients with Familial Hypercholesterolemia [25, 26]. Figure 5.3, shows the calcification in the aorta.

The Bone-Heart Paradox: Atherosclerosis in the Heart and in the Bone

Atherosclerosis and osteoporosis are common medical conditions, which increases in prevalence with the aging population. Recent studies are demonstrate parallel risk factors for the development of these disease processes [27]; however, the phenotypic expression of experimental hypercholesterolemia in the aortic valves and femurs were recently defined in the LDLR null mouse model [24].

Atherosclerosis is characterized as complex inflammatory disease which develops secondary to risk factors such as elevated cholesterol, hypertension, smoking, male gender and postmenopausal women [28–30]. The initial atherosclerotic event has been characterized as the fatty streak lesion, which represents a complex series of signaling events and inflammatory cells accumulating along the vascular and valvular surfaces [3, 23, 31–40]. There are increasing number of studies which correlate osteoporosis with cardiovascular risk factors [27, 41, 42].

Osteoporosis continues to be the leading cause of bone fractures in postmenopausal women, leading to a high morbidity and mortality [28, 43, 44]. There is increasing evidence that both of these disease processes develop in parallel, with an associated phenotype of decrease bone formation in the bones and increased bone formation in the cardiovascular system [44]. For years lipids have been hypothesized as a mechanism of both disease processes but cellular processes are not well known.

Cardiovascular calcification and osteoporosis are the most common causes of morbidity and mortality in the USA. The spectrum of degenerative valve lesions have traditionally been thought to be due to a passive disease process developing rapidly within the valve leaflets. The most common location of calcific aortic valve disease is the left side of the heart [23, 45]. Recently, the biology of the heart valve has changed from degeneration of the valve to an active biologic disease process in the heart valve [46]. Despite the high incidence of valvular heart disease, the signaling pathways in human valve disease are still under intense investigation [3, 5, 23, 33, 45–47]. Understanding the parallel role of bone in the heart and femurs is becoming increasing important, since the phenotype of calcification in the valve has a similar osteogenic mechanism to that of endochondral bone formation [3].

The roles of lipids in regulating the osteogenic mechanism via the Lrp5/Wnt pathway are evolving [48, 49]. These studies are providing more evidence for the lipid hypothesis in bones as well as the heart. Previously studies have shown in the experimental hypercholesterolemic rabbit model that the aortic valve expresses atherosclerosis, calcification, attenuation of eNOS expression, regulation of Cbfa1 and upregulation of Lrp5 receptor in the calcifying aortic valve [4, 38, 50–52]. Recently, a similar effect

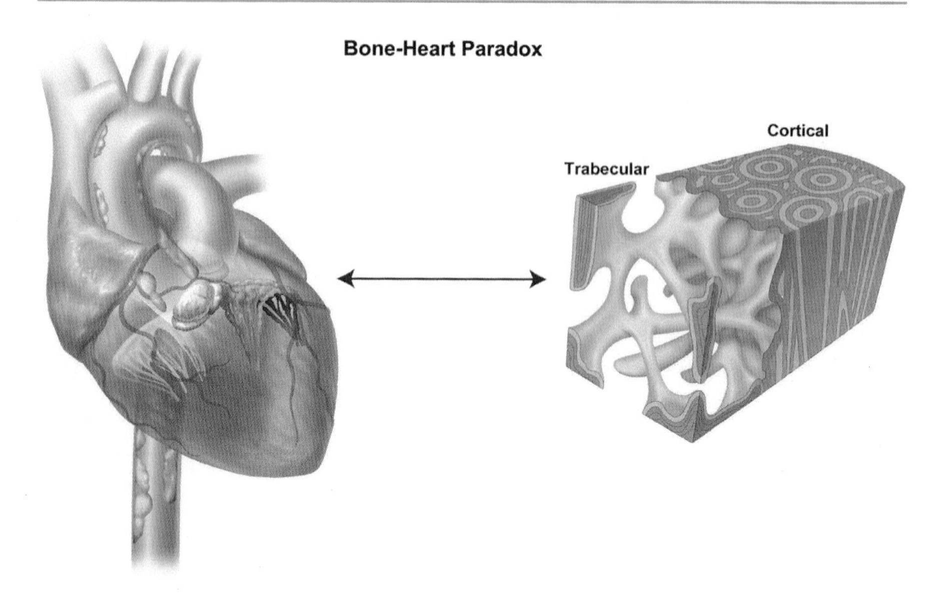

Fig. 5.4 The bone paradox demonstrates the calcification in the heart and the osteoporosis in the bone

was shown using molecular imaging in the eNOS$^{-/-}$ mouse model of experimental hypercholesterolemia [49].

Studies have demonstrated in the past, that vascular smooth muscle cell is responsible for the calcifying phenotype within the vessel [7, 53–61] and has developed a similar hypothesis in the vasculature. Our laboratory has shown a similar effect in the aortic valve in the presence of hypercholesterolemia [4, 36, 37, 40, 49, 62]. A recent report determines if there is an increase in mineralization in the heart and a decrease in the skeleton, using an experimental hypercholesterolemic LDLR$^{-/-}$ mouse model with and without atorvastatin [24]. This study is the first to test *in vivo* the effects of a lipid diet in the LDLR$^{-/-}$ mouse model. Figure 5.4, shows the atherosclerotic mechanism in the bone and in the heart.

The LDLR$^{-/-}$ mice develop similar findings using histology, calcein incorporation and MicroCT as the valves calcify the bones develop decrease mineralization in the presence of elevated cholesterol diet. This study also demonstrates that calcein incorporation in the valves and the bones of the hypercholesterolemic model indicates an active process of bone turnover with increases in macrophage cells in both the valve and femur. The MicroCT results demonstrate that as the valves mineralize the bones develop less mineralization. The phenotypic characteristics of the hypercholesterolemic valves and femurs demonstrate (1) increase in macrophage cells in the valves and the femurs, (2) increase in calcein incorporation in the valves and the femurs, (3) increase in calcification in the valves and decrease in the calcification in the femurs.

The foam cell lesion along the surface of the aortic valve leaflet, is similar to previously published data of foam cells accumulating in the atherosclerotic aortic valve in the presence of cholesterol diet [34, 36, 39, 40].These histologic findings demonstrate the potential atherosclerotic connection between these two disease processes which includes cellular activation, lipid accumulation, calcein incorporation and change in the mineralization process simultaneously in the two different target organs. The change in mineralization is the critical paradox in the field of atherosclerosis and osteoporosis. The movement of the calcium mineral from the femurs to the heart has been hypothesized over the last decades. The role of statins reverses this paradox, potentially via Wnt/Lrp5 [38] mechanism, versus BMP [63, 64] mechanisms. Further studies testing the roles of HMG CoA reductase agents are critical for the future therapy for this patient population.

This mouse model is the provide evidence for this paradoxical role of bone in the aortic valve and the femur in an experimental hypercholesterolemic LDLR$^{-/-}$ mouse model. The presence of elevated cholesterol induces inflammatory cells to infiltrate and activate the classical histologic findings of atherosclerosis in the heart valve. In the femur the presence of decrease bone formation by MicroCT, active bone turnover, and the presence of macrophage cells indicates a parallel process in this organ. As the mineralization process is increased in the heart, it is decreased in the femurs, which chronically leads to progressive decrease in the bone thickness and over time potential osteoporotic changes, simultaneously causing aortic valve stenosis.

Summary

Endochondral bone formation in the heart and in the bone has parallel cellular mechanisms. Risk factors for the development of atherosclerosis in the bone and in the heart are similar and contribute to the bone-heart paradox. The clinical manifestations of these two disease processes are long-term osteoporosis and cardiovascular calcification. Identifying the role of osteocardiology risk factors and subclinical disease will in the future help to identify therapeutic approaches to slow the progression of these disease processes in the future.

References

1. O'Brien KD, Kuusisto J, Reichenbach DD, et al. Osteopontin is expressed in human aortic valvular lesions [comment]. Circulation. 1995;92:2163–8.
2. Mohler ER 3rd, Gannon F, Reynolds C, Zimmerman R, Keane MG, Kaplan FS. Bone formation and inflammation in cardiac valves. Circulation. 2001;103:1522–8.
3. Rajamannan NM, Subramaniam M, Rickard D, et al. Human aortic valve calcification is associated with an osteoblast phenotype. Circulation. 2003;107:2181–4.
4. Rajamannan NM, Subramaniam M, Springett M, et al. Atorvastatin inhibits hypercholesterolemia-induced cellular proliferation and bone matrix production in the rabbit aortic valve. Circulation. 2002;105:2260–5.
5. Caira FC, Stock SR, Gleason TG, et al. Human degenerative valve disease is associated with up-regulation of low-density lipoprotein receptor-related protein 5 receptor-mediated bone formation. J Am Coll Cardiol. 2006;47:1707–12.
6. Jian B, Jones PL, Li Q, Mohler ER 3rd, Schoen FJ, Levy RJ. Matrix metalloproteinase-2 is associated with tenascin-C in calcific aortic stenosis. Am J Pathol. 2001;159:321–7.
7. Tintut Y, Alfonso Z, Saini T, et al. Multilineage potential of cells from the artery wall. Circulation. 2003;108:2505–10.

8. Parhami F, Basseri B, Hwang J, Tintut Y, Demer LL. High-density lipoprotein regulates calcification of vascular cells. Circ Res. 2002;91:570–6.
9. Parhami F, Morrow AD, Balucan J, et al. Lipid oxidation products have opposite effects on calcifying vascular cell and bone cell differentiation. A possible explanation for the paradox of arterial calcification in osteoporotic patients. Arterioscler Thromb Vasc Biol. 1997;17:680–7.
10. Mohler ER 3rd, Adam LP, McClelland P, Graham L, Hathaway DR. Detection of osteopontin in calcified human aortic valves. Arterioscler Thromb Vasc Biol. 1997;17:547–52.
11. O'Brien KD, Kuusisto J, Reichenbach DD, et al. Osteopontin is expressed in human aortic valvular lesions. Circulation. 1995;92:2163–8.
12. Shao JS, Cheng SL, Pingsterhaus JM, Charlton-Kachigian N, Loewy AP, Towler DA. Msx2 promotes cardiovascular calcification by activating paracrine Wnt signals. J Clin Invest. 2005;115:1210–20.
13. Tintut Y, Parhami F, Bostrom K, Jackson SM, Demer LL. cAMP stimulates osteoblast-like differentiation of calcifying vascular cells. Potential signaling pathway for vascular calcification. J Biol Chem. 1998;273:7547–53.
14. Tintut Y, Patel J, Territo M, Saini T, Parhami F, Demer LL. Monocyte/macrophage regulation of vascular calcification in vitro. Circulation. 2002;105:650–5.
15. Demer LL, Tintut Y. Inflammatory, metabolic, and genetic mechanisms of vascular calcification. Arterioscler Thromb Vasc Biol. 2014;34:715–23.
16. Boyden LM, Mao J, Belsky J, et al. High bone density due to a mutation in LDL-receptor-related protein 5. N Engl J Med. 2002;346:1513–21.
17. Gong Y, Slee RB, Fukai N, et al. LDL receptor-related protein 5 (LRP5) affects bone accrual and eye development. Cell. 2001;107:513–23.
18. Fujino T, Asaba H, Kang MJ, et al. Low-density lipoprotein receptor-related protein 5 (LRP5) is essential for normal cholesterol metabolism and glucose-induced insulin secretion. Proc Natl Acad Sci U S Am. 2003;100:229–34.
19. Babij P, Zhao W, Small C, et al. High bone mass in mice expressing a mutant LRP5 gene. J Bone Miner Res. 2003;18:960–74.
20. Westendorf JJ, Kahler RA, Schroeder TM. Wnt signaling in osteoblasts and bone diseases. Gene. 2004;341:19–39.
21. Holmen SL, Giambernardi TA, Zylstra CR, et al. Decreased BMD and limb deformities in mice carrying mutations in both Lrp5 and Lrp6. J Bone Miner Res. 2004;19:2033–40.
22. Awan Z, Denis M, Bailey D, et al. The LDLR deficient mouse as a model for aortic calcification and quantification by micro-computed tomography. Atherosclerosis. 2011;219:455–62.
23. Rajamannan NM. The role of Lrp5/6 in cardiac valve disease: experimental hypercholesterolemia in the ApoE−/− /Lrp5−/− mice. J Cell Biochem. 2011;112:2987–91.
24. Rajamannan NM. Atorvastatin attenuates bone loss and aortic valve atheroma in LDLR mice. Cardiology. 2015;132:11–5.
25. Rajamannan NM. Calcific aortic valve disease in familial hypercholesterolemia: the LDL-density-gene effect. J Am Coll Cardiol. 2015;66:2696–8.
26. ten Kate GJ, Bos S, Dedic A, et al. Increased aortic valve calcification in familial hypercholesterolemia: prevalence, extent, and associated risk factors. J Am Coll Cardiol. 2015;66:2687–95.
27. Figueiredo CP, Rajamannan NM, Lopes JB, et al. Serum phosphate and hip bone mineral density as additional factors for high vascular calcification scores in a community-dwelling: The Sao Paulo Ageing & Health Study (SPAH). Bone. 2012;52:354–9.
28. Brochier ML, Arwidson P. Coronary heart disease risk factors in women. Eur Heart J. 1998;19(Suppl A):A45–52.
29. Montalcini T, Gorgone G, Pujia A. Association between pulse pressure and subclinical carotid atherosclerosis in normotensive and hypertensive post-menopausal women. Clin Exp Hypertens. 2009;31:64–70.
30. Bolego C, Poli A, Paoletti R. Smoking and gender. Cardiovasc Res. 2002;53:568–76.
31. Rajamannan NM. Calcific aortic stenosis: a disease ready for prime time. Circulation. 2006;114:2007–9.
32. Rajamannan NM. Low-density lipoprotein and aortic stenosis. Heart. 2008;94:1111–2.

33. Rajamannan NM. Mechanisms of aortic valve calcification: the LDL-density-radius theory A: translation from cell signaling to physiology. Am J Physiol. 2010;298:H5–15.
34. Rajamannan NM. Oxidative-mechanical stress signals stem cell niche mediated Lrp5 osteogenesis in eNOS(−/−) null mice. J Cell Biochem. 2012;113:1623–34.
35. Rajamannan NM, Bonow RO, Rahimtoola SH. Calcific aortic stenosis: an update. Nat Clin Pract. 2007;4:254–62.
36. Rajamannan NM, Edwards WD, Spelsberg TC. Hypercholesterolemic aortic-valve disease. N Engl J Med. 2003;349:717–8.
37. Rajamannan NM, Sangiorgi G, Springett M, et al. Experimental hypercholesterolemia induces apoptosis in the aortic valve. J Heart Valve Dis. 2001;10:371–4.
38. Rajamannan NM, Subramaniam M, Caira F, Stock SR, Spelsberg TC. Atorvastatin inhibits hypercholesterolemia-induced calcification in the aortic valves via the Lrp5 receptor pathway. Circulation. 2005;112:I229–34.
39. Rajamannan NM, Subramaniam M, Springett M, et al. Atorvastatin inhibits hypercholesterolemia-induced cellular proliferation and bone matrix production in the rabbit aortic valve. Circulation. 2002;105:2660–5.
40. Rajamannan NM, Subramaniam M, Stock SR, et al. Atorvastatin inhibits calcification and enhances nitric oxide synthase production in the hypercholesterolaemic aortic valve. Heart. 2005;91:806–10.
41. Burnett JR, Vasikaran SD. Cardiovascular disease and osteoporosis: is there a link between lipids and bone? Ann Clin Biochem. 2002;39:203–10.
42. Hjortnaes J, Butcher J, Figueiredo JL, et al. Arterial and aortic valve calcification inversely correlates with osteoporotic bone remodelling: a role for inflammation. Eur Heart J. 2010;31:1975–84.
43. van der Schouw YT, Grobbee DE. Menopausal complaints, oestrogens, and heart disease risk: an explanation for discrepant findings on the benefits of post-menopausal hormone therapy. Eur Heart J. 2005;26:1358–61.
44. Tekin GO, Kekilli E, Yagmur J, et al. Evaluation of cardiovascular risk factors and bone mineral density in post menopausal women undergoing coronary angiography. Int J Cardiol. 2008;131:66–9.
45. Rajamannan NM. The role of Lrp5/6 in cardiac valve disease: LDL-density-pressure theory. J Cell Biochem. 2011;112:2222–9.
46. Rajamannan NM, Evans FJ, Aikawa E, et al. Calcific aortic valve disease: not simply a degenerative process: a review and agenda for research from the National Heart and Lung and Blood Institute Aortic Stenosis Working Group. Executive summary: calcific aortic valve disease-2011 update. Circulation. 2011;124:1783–91.
47. Rajamannan NM. Calcific aortic valve disease: cellular origins of valve calcification. Arterioscler Thromb Vasc Biol. 2011;31:2777–8.
48. Almeida M, Ambrogini E, Han L, Manolagas SC, Jilka RL. Increased lipid oxidation causes oxidative stress, increased peroxisome proliferator-activated receptor-gamma expression, and diminished pro-osteogenic Wnt signaling in the skeleton. J Biol Chem. 2009;284:27438–48.
49. Rajamannan NM. Oxidative-mechanical stress signals stem cell niche mediated Lrp5 osteogenesis in eNOS(−/−) null mice. J Cell Biochem. 2012;113:1623–34.
50. Rajamannan NM, Caplice N, Anthikad F, et al. Cell proliferation in carcinoid valve disease: a mechanism for serotonin effects. J Heart Valve Dis. 2001;10:827–31.
51. Rajamannan NM. Calcific aortic stenosis: medical and surgical management in the elderly. Curr Treat Options Cardiovasc Med. 2005;7:437–42.
52. Rajamannan NM, Nealis TB, Subramaniam M, et al. Calcified rheumatic valve neoangiogenesis is associated with vascular endothelial growth factor expression and osteoblast-like bone formation. Circulation. 2005;111:3296–301.
53. Tintut Y, Demer LL. Recent advances in multifactorial regulation of vascular calcification. Curr Opin Lipidol. 2001;12:555–60.
54. Abedin M, Tintut Y, Demer LL. Vascular calcification: mechanisms and clinical ramifications. Arterioscler Thromb Vasc Biol. 2004;24:1161–70.

55. Tintut Y, Abedin M, Cho J, Choe A, Lim J, Demer LL. Regulation of RANKL-induced osteoclastic differentiation by vascular cells. J Mol Cell Cardiol. 2005;39:389–93.
56. Abedin M, Tintut Y, Demer LL. Mesenchymal stem cells and the artery wall. Circ Res. 2004;95:671–6.
57. Garfinkel A, Tintut Y, Petrasek D, Bostrom K, Demer LL. Pattern formation by vascular mesenchymal cells. Proc Natl Acad Sci U S A. 2004;101:9247–50.
58. Tintut Y, Morony S, Demer LL. Hyperlipidemia promotes osteoclastic potential of bone marrow cells ex vivo. Arterioscler Thromb Vasc Biol. 2004;24:e6–10.
59. Mody N, Tintut Y, Radcliff K, Demer LL. Vascular calcification and its relation to bone calcification: possible underlying mechanisms. J Nucl Cardiol. 2003;10:177–83.
60. Parhami F, Tintut Y, Beamer WG, Gharavi N, Goodman W, Demer LL. Atherogenic high-fat diet reduces bone mineralization in mice. J Bone Miner Res. 2001;16:182–8.
61. Parhami F, Tintut Y, Patel JK, Mody N, Hemmat A, Demer LL. Regulation of vascular calcification in atherosclerosis. Z Kardiol. 2001;90(Suppl 3):27–30.
62. Makkena B, Salti H, Subramaniam M, et al. Atorvastatin decreases cellular proliferation and bone matrix expression in the hypercholesterolemic mitral valve. J Am Coll Cardiol. 2005;45:631–3.
63. Kupcsik L, Meurya T, Flury M, Stoddart M, Alini M. Statin-induced calcification in human mesenchymal stem cells is cell death related. J Cell Mol Med. 2009;13:4465–73.
64. Maeda T, Matsunuma A, Kurahashi I, Yanagawa T, Yoshida H, Horiuchi N. Induction of osteoblast differentiation indices by statins in MC3T3-E1 cells. J Cell Biochem. 2004;92:458–71.

Osteocardiology: The Atherosclerotic Bone Paradox

6

Introduction

The discovery of the Lrp5 receptor in the gain-of-function [6], and loss-of-function [7] mutations in bone diseases led to several studies showing that activation of the canonical Wnt pathway is important in osteoblastogenesis [5, 8, 9]. For years, the signaling mechanisms important on osteogenesis include BMP, TGFβ, SMAD signaling, etc., which have led the field in determining bone formation and osteoporosis, however, the discovery of Wnt Signaling in bone homeostasis has become a major mechanism in osteogenesis in the heart and in the bone. The bone-heart paradox, has helped to identify specific genes that may contribute to the development of osteoporosis and cardiovascular calcification through the use of transgenic and knockout mouse models. These studies have identified a handful of master genes, including Runx2 and osterix, which are absolutely necessary for ossification as deletion of these genes in mice completely ablates bone formation [1, 2]. An additional set of genes, known as modulators of skeletal development and bone homeostasis, have also been revealed which in some cases regulate the activity of these master genes, including Sox9 and Msx2. Wnt signaling activates these master genes from the extracellular membrane activation of the trimeric complex which includes: Lrp5, Wnt, and Frizzled. Downstream of this complex, in the nucleus, a series of steps regulates transcriptional activation of the osteogenic cascade.

Recently, TGFβ Inducible Early Gene-1 (TIEG) [3] has played a crucial roles for TIEG1 in regulating Wnt Signaling a wide variety of cellular processes and molecular functions important for bone biology. These include modulation of the TGFβ, BMP, and estrogen signaling pathways [4–15], osteoblast and osteoclast functions [16–19] as well as regulation of skeletal development and homeostasis [12, 20, 21]. Recently, allelic variations in the TIEG1 gene [22] and altered TIEG1 expression levels [23] have been identified in patients with osteoporosis. These studies, have implicated a central role for TIEG1 in mediating bone growth and homeostasis.

N.M. Rajamannan, *Osteocardiology*, DOI 10.1007/978-3-319-64994-8_6

Ongoing studies have now revealed an important role for TIEG1 in regulating the activity of the canonical Wnt pathway in bone. This is of significant interest given that Wnt signaling is essential for bone formation and bone related diseases. Recently, the role of TIEG1 in downstream regulation of Wnt signaling at the level of transcription in bone [24] and in the heart has been established.

The Role of TIEG1 in Wnt Signaling in Bone

A recent study demonstrates that loss of TIEG1 expression is associated with alterations in the expression levels of multiple Wnt ligands and downstream mediators of the pathway resulting in suppression of Wnt signaling in osteoblast cells and throughout the mouse skeleton [24]. TIEG1 is shown to enhance Wnt signaling through at least two different mechanisms; one by suppressing GSK-3β activity and inducing β-catenin nuclear localization, and two by serving as a co-activator for Lef1 and β-catenin transcriptional activity. The data link the critical role for TIEG1 in mediating cross-talk between the TGFβ/BMP and Wnt signaling pathways. Given the importance of Wnt signaling for skeletal development and bone homeostasis, and the present data linking a role for TIEG1 in mediating this pathway, it is likely that alterations in Wnt signaling contribute to the observed osteopenic phenotype of TIEG1 KO mice. It is possible that future therapies would also be relevant for the treatment of osteoporosis in individuals with TIEG1 polymorphisms or altered TIEG1 expression levels as has been previously reported.

The Role of TIEG1 in Wnt Signaling in the Heart

Atherosclerosis and osteoporosis are common medical conditions, which are increasing in prevalence as the population is aging throughout the world. Recently, studies demonstrate that atherosclerosis is present in hyperlipidemic bones and valves as characterized by macrophage and osteoclast infiltration, which is attenuated by atorvastatin [25]. A recent report tested the role of Wnt Signaling in the $LDLR^{-/-}$ mouse model, which is a clinical surrogate for familial hypercholesterolemia the genetic disorder of lipid abnormalities in patients as related to lack of functioning LDLR receptor. The study confirmed that LiCl, a known activator of Wnt Signaling is active in the development of atherosclerosis in the heart and in the femur [26].

For decades, CAVD was thought to be due to degenerative process [27]. Studies demonstrate that the Wnt pathway plays important roles in valve calcification associated with a specific osteogenic phenotype [28] defined by increased bone mineral content. Furthermore, recent studies demonstrate the role of Wnt Signaling in the development of calcific aortic valve disease including the role of Lrp5 and Lrp6, which both play a role in the mineralizing valve secondary to chronic hypercholesterolemia. [29–33] The most recent discovery of the role of TIEG1, a transcription factor known to play critical roles in osteoblast differentiation and bone mineralization bone

[3, 34], is also be involved in mediating the Wnt signaling pathway in the skeleton and in the heart [24].

Finally, the role of TIEG1 is also critical in the transcriptional regulation of Wnt signaling in aortic valve interstitial cells, which provides proof of principle that the atherosclerotic hypothesis of Wnt Signaling in the aortic valve is mediated via TIEG1 signaling down stream in the nucleus. To further understand the mechanisms of Wnt signaling in the development of calcification in the aortic valve, the recent discovery that TIEG1 enhances Wnt signaling by regulating β-catenin nuclear localization and by serving as a co-activator for Lef1 and β-catenin transcriptional activity [24]. Known activators of Wnt signaling, Wnt3a, TGF β and BMP-4 a member of the TGF superfamily demonstrated VIC proliferation and VIC alkaline phosphatase synthesis [26]. This is the first evidence to demonstrate the role of TIEG1 signaling in the cardiac valve [26] and in the bone [24]. Taken together, these data implicate an important role for TIEG1 in mediating Wnt signaling and LEF transcriptional activity in VICs. In the future, this signaling pathway may be a potential target for medical therapy in the future to slow the progression of CAVD [27].

Wnt Signaling in Atherosclerosis

The low density lipoprotein co-receptor Lrp5/6 is a member of the family of structurally closely related cell surface low density lipoprotein receptors that have diverse biological functions in different organs, tissues and cell types which are important in development and disease mechanisms. The most prominent role in this evolutionary ancient family is cholesterol homeostasis. The LRP5 pathway regulates bone formation in different diseases of bone [35, 36]. The discovery of the LRP5 receptor in the gain of function [36] and loss of function [35] mutations in the development of bone diseases, resulted in a number of studies which have shown that activation of the canonical Wnt pathway is important in osteoblastogenesis [37–40]. Three studies to date have confirmed the regulation of the LRP5/Wnt pathway for cardiovascular calcification *in vivo* and *ex vivo* [29–31]. Lrp5 has been shown to have an effect on bone mass via the mechanostat effect on regulating bone formation. The findings in the human of the high bone mass gain of function mutation [41], led to a series of discoveries that Lrp5 regulates bone mass via the mechanical force effect on the receptor [42–44]. Lrp6 also regulates bone but has been found to have a low bone mass effect in patients in which a putative partial loss-of-function mutation in LRP6 was identified to early cardiovascular-related death associated with increased plasma LDL, triglycerides, hypertension, diabetes and osteoporosis [45]. This data is the first to demonstrate in the genetic mice to demonstrate that experimental cholesterol diet can upregulate Lrp5 and Lrp6 with varying degrees of calcification. The $ApoE^{-/-}$ demonstrated marked increase in the calcification which is consistent with the lipid and pressure effect of Lrp5 on the aortic valves. The $Lrp5^{-/-}$ had no calcification in the valves. The $Lrp5^{-/-}$ single gene KO demonstrates the role of Lrp5 for calcification and the $ApoE^{-/-}$ single gene knockout to demonstrate the role cholesterol to activate the Lrp5/6 receptors. The double knockout mice ApoE$^{-/-}$: Lrp5$^{-/-}$ were tested to show that in the elevated lipids secondary to the lack of the ApoE receptor as compared to the Lrp5$^{-/-}$ mice caused some mild calcification via the upregulation of the Lrp6 gene expression in the mice (Fig. 6.1).

The right-sided valves in all of the specific mice did not develop any calcification, which further demonstrates the role of the higher pressures in the left side of the heart to activate the Lrp5/6 receptor in the valve. LRP5 binds apoE-containing lipoproteins *in vitro* and is widely expressed in many tissues including hepatocytes, adrenal gland and pancreas [46]. The production of mice lacking LRP5 revealed that LRP5 deficiency led to increased plasma cholesterol levels in mice fed a high-fat diet, secondary to decreased hepatic clearance of chylomicron remnants and also

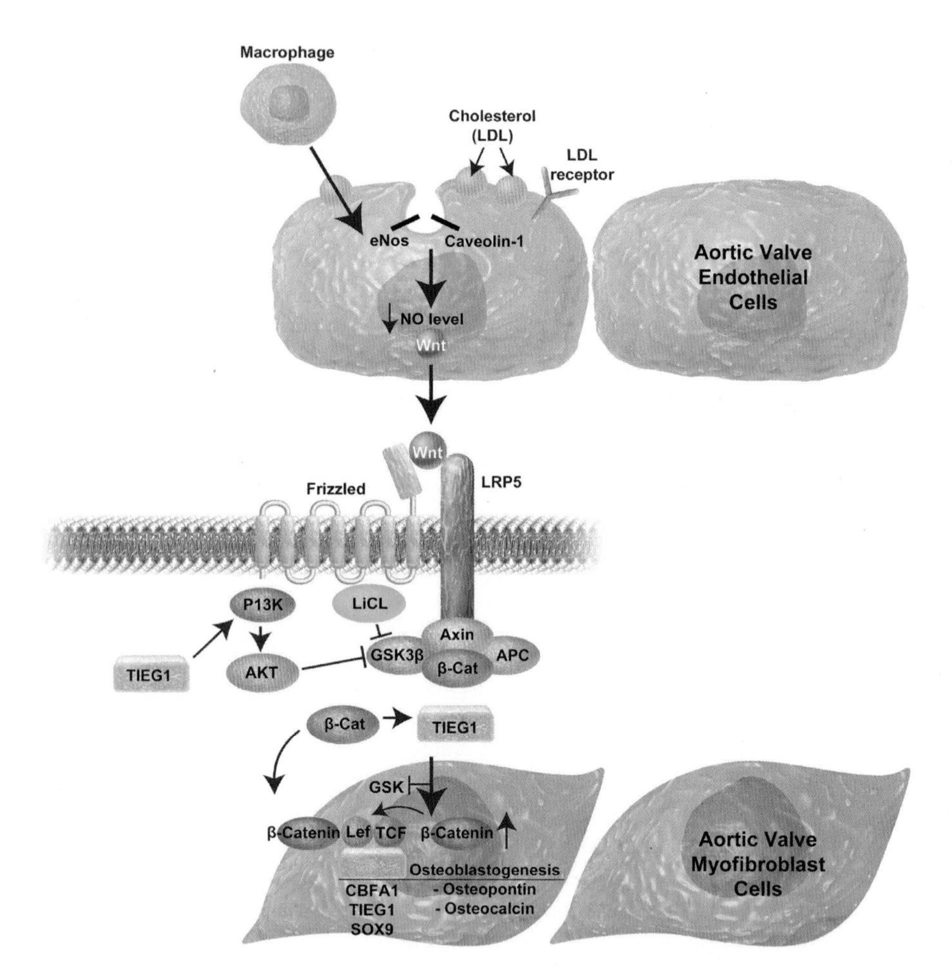

Fig. 6.1 Wnt Signaling in Atherosclerosis: Osteocardiology, the presence of LiCl, GSK9 is inhibited and β-Catenin translocates to the nucleus to activate osteogenesis via TIEG1, Cbfa1 and LEF/TCF in the valve and in the femur [26]

marked impaired glucose tolerance [37]. In the LRP5 mice that were not fed the high cholesterol diet, the mice did not develop high cholesterol levels [47]. The investigators went on to define the role of LRP5 in the lipoprotein metabolism by developing a double knockout mouse for ApoE:LRP5. They found that the double KO mouse had approx 60% higher cholesterol levels compared with the age matched apoE knockout mice. High performance liquid chromatography analysis of plasma lipoproteins revealed that no difference in the apoproteins but the cholesterol levels in the very low density and low density lipoprotein fractions were markedly increased in the apoE:Lrp5 double KO mice. There was threefold increase in the atherosclerosis indicating that the Lrp5 mediates both apoE-dependent and apoE-independent catabolism of lipoproteins. In this current study performed the serum cholesterol levels and demonstrated marked increase in the cholesterol in both of the ApoE$^{-/-}$ and the ApoE:Lrp5 double KO mice further confirming the association of elevated cholesterol and the mineralization process.

In Summary

Atherosclerotic mechanisms of the bone-heart paradox, in the presence of osteocardiology risk factors atherosclerosis ensues in the heart and in the bone to develop calcification in the heart and osteoporosis in the bone as shown in Fig. 6.2, in which lipids and mechanical pressure activate Lrp5/Wnt signaling in the heart and in the bone to initiate the atherosclerotic bone paradox.

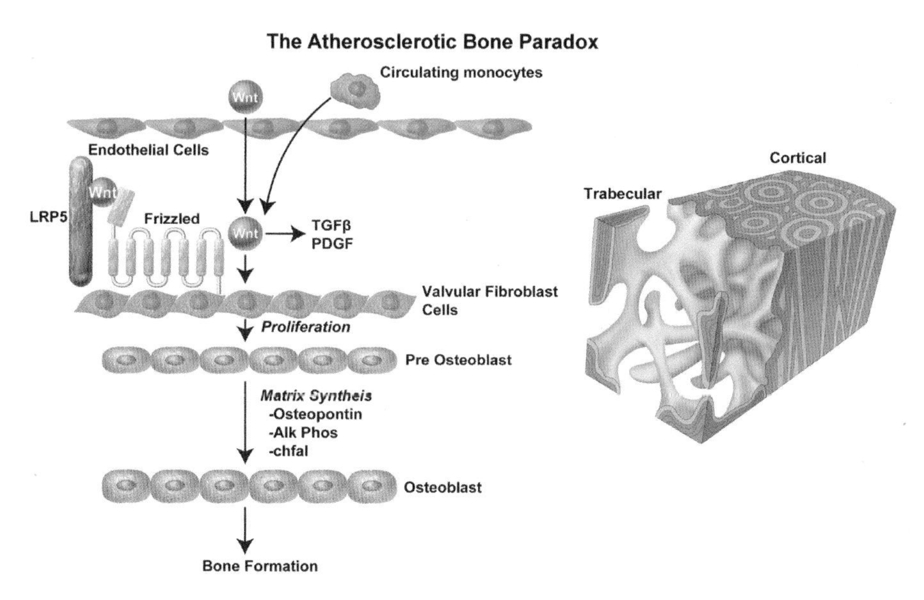

Fig. 6.2 Atherosclerotic mechanisms of the bone-heart paradox, in the presence of osteocardiology risk factors atherosclerosis ensues in the heart and in the bone to develop calcification in the heart and osteoporosis in the bone

References

1. Komori T, Yagi H, Nomura S, et al. Targeted disruption of Cbfa1 results in a complete lack of bone formation owing to maturational arrest of osteoblasts. Cell. 1997;89:755–64.
2. Nakashima K, Zhou X, Kunkel G, et al. The novel zinc finger-containing transcription factor osterix is required for osteoblast differentiation and bone formation. Cell. 2002;108:17–29.
3. Subramaniam M, Harris SA, Oursler MJ, Rasmussen K, Riggs BL, Spelsberg TC. Identification of a novel TGF-beta-regulated gene encoding a putative zinc finger protein in human osteoblasts. Nucleic Acids Res. 1995;23:4907–12.
4. Hefferan TE, Reinholz GG, Rickard DJ, et al. Overexpression of a nuclear protein, TIEG, mimics transforming growth factor-beta action in human osteoblast cells. J Biol Chem. 2000;275:20255–9.
5. Tachibana I, Imoto M, Adjei PN, et al. Overexpression of the TGFbeta-regulated zinc finger encoding gene, TIEG, induces apoptosis in pancreatic epithelial cells. J Clin Invest. 1997;99:2365–74.
6. Johnsen SA, Subramaniam M, Janknecht R, Spelsberg TC. TGFbeta inducible early gene enhances TGFbeta/Smad-dependent transcriptional responses. Oncogene. 2002;21: 5783–90.
7. Hefferan TE, Subramaniam M, Khosla S, Riggs BL, Spelsberg TC. Cytokine-specific induction of the TGF-beta inducible early gene (TIEG): regulation by specific members of the TGF-beta family. J Cell Biochem. 2000;78:380–90.
8. Bensamoun SF, Tsubone T, Subramaniam M, et al. Age-dependent changes in the mechanical properties of tail tendons in TGF-beta inducible early gene-1 knockout mice. J Appl Physiol. 2006;101:1419–24.
9. Johnsen SA, Subramaniam M, Katagiri T, Janknecht R, Spelsberg TC. Transcriptional regulation of Smad2 is required for enhancement of TGFbeta/Smad signaling by TGFbeta inducible early gene. J Cell Biochem. 2002;87:233–41.
10. Johnsen SA, Subramaniam M, Monroe DG, Janknecht R, Spelsberg TC. Modulation of transforming growth factor beta (TGFbeta)/Smad transcriptional responses through targeted degradation of TGFbeta-inducible early gene-1 by human seven in absentia homologue. J Biol Chem. 2002;277:30754–9.
11. Tsubone T, Moran SL, Subramaniam M, Amadio PC, Spelsberg TC, An KN. Effect of TGF-beta inducible early gene deficiency on flexor tendon healing. J Orthop Res. 2006;24:569–75.
12. Hawse J, Iwaniec UT, Bensamoun SF, Monroe DG, Peters KD, Ilharreborde B, Rajamannan NM, Oursler MJ, Turner RT, Spelsberg TC, Subramaniam M. TIEG-Null mice display an osteopenic gender-specific phenotype. Bone. 2008;42:1025–31.
13. Hawse JR, Subramaniam M, Monroe DG, et al. Estrogen receptor beta isoform-specific induction of transforming growth factor beta-inducible early gene-1 in human osteoblast cells: an essential role for the activation function 1 domain. Mol Endocrinol (Baltimore, Md). 2008;22:1579–95.
14. Taguchi M, Moran SL, Zobitz ME, et al. Wound-healing properties of transforming growth factor beta (TGF-beta) inducible early gene 1 (TIEG1) knockout mice. J Musculoskelet Res. 2008;11:63–9.
15. Venuprasad K, Huang H, Harada Y, et al. The E3 ubiquitin ligase Itch regulates expression of transcription factor Foxp3 and airway inflammation by enhancing the function of transcription factor TIEG1. Nat Immunol. 2008;9:245–53.
16. Subramaniam M, Gorny G, Johnsen SA, et al. TIEG1 null mouse-derived osteoblasts are defective in mineralization and in support of osteoclast differentiation in vitro. Mol Cell Biol. 2005;25:1191–9.
17. Subramaniam M, Hawse JR, Bruinsma ES, et al. TGFbeta inducible early gene-1 directly binds to, and represses, the OPG promoter in osteoblasts. Biochem Biophys Res Commun. 2010;392:72–6.
18. Subramaniam M, Hawse JR, Rajamannan NM, Ingle JN, Spelsberg TC. Functional role of KLF10 in multiple disease processes. BioFactors (Oxford, England). 2010;36:8–18.

19. Cicek M, Vrabel A, Sturchio C, et al. TGF-beta inducible early gene 1 regulates osteoclast differentiation and survival by mediating the NFATc1, AKT, and MEK/ERK signaling pathways. PLoS One. 2011;6:e17522.
20. Bensamoun SF, Hawse JR, Subramaniam M, et al. TGFbeta inducible early gene-1 knockout mice display defects in bone strength and microarchitecture. Bone. 2006;39:1244–51.
21. Subramaniam M, Hawse JR, Johnsen SA, Spelsberg TC. Role of TIEG1 in biological processes and disease states. J Cell Biochem. 2007;102:539–48.
22. Yerges LM, Klei L, Cauley JA, et al. Candidate gene analysis of femoral neck trabecular and cortical volumetric bone mineral density in older men. J Bone Miner Res. 2010;25:330–8.
23. Hopwood B, Tsykin A, Findlay DM, Fazzalari NL. Gene expression profile of the bone microenvironment in human fragility fracture bone. Bone. 2009;44:87–101.
24. Subramaniam M, Cicek M, Pitel KS, et al. TIEG1 modulates beta-catenin sub-cellular localization and enhances Wnt signaling in bone. Nucleic Acids Res. 2017;45:5170–82.
25. Rajamannan NM. Atorvastatin attenuates bone loss and aortic valve atheroma in LDLR mice. Cardiology. 2015;132:11–5.
26. Rajamannan NM. The role of TIEG1 in calcific aortic valve disease. Journal of Bone and Mineral Metabolism. 2017;29(9):S136.
27. Rajamannan NM. Evans FJ, Aikawa E, et al. Calcific aortic valve disease: not simply a degenerative process: a review and agenda for research from the National Heart and Lung and Blood Institute Aortic Stenosis Working Group. Executive summary: Calcific aortic valve disease-2011 update. Circulation. 2011;124:1783–91.
28. Rajamannan NM, Subramaniam M, Rickard D, et al. Human aortic valve calcification is associated with an osteoblast phenotype. Circulation. 2003;107:2181–4.
29. Caira FC, Stock SR, Gleason TG, et al. Human degenerative valve disease is associated with up-regulation of low-density lipoprotein receptor-related protein 5 receptor-mediated bone formation. J Am Coll Cardiol. 2006;47:1707–12.
30. Rajamannan NM, Subramaniam M, Caira F, Stock SR, Spelsberg TC. Atorvastatin inhibits hypercholesterolemia-induced calcification in the aortic valves via the Lrp5 receptor pathway. Circulation. 2005;112:I229–34.
31. Shao JS, Cheng SL, Pingsterhaus JM, Charlton-Kachigian N, Loewy AP, Towler DA. Msx2 promotes cardiovascular calcification by activating paracrine Wnt signals. J Clin Invest. 2005;115:1210–20.
32. Rajamannan NM. The role of Lrp5/6 in cardiac valve disease: experimental hypercholesterolemia in the ApoE−/− /Lrp5−/− mice. J Cell Biochem. 2011;112:2987–91.
33. Rajamannan NM. Oxidative-mechanical stress signals stem cell niche mediated Lrp5 osteogenesis in eNOS(−/−) null mice. J Cell Biochem. 2012;113:1623–34.
34. Hawse JR, Cicek M, Grygo SB, et al. TIEG1/KLF10 modulates Runx2 expression and activity in osteoblasts. PLoS One. 2011;6:e19429.
35. Gong Y, Slee RB, Fukai N, et al. LDL receptor-related protein 5 (LRP5) affects bone accrual and eye development. Cell. 2001;107:513–23.
36. Boyden LM, Mao J, Belsky J, et al. High bone density due to a mutation in LDL-receptor-related protein 5. N Engl J Med. 2002;346:1513–21.
37. Fujino T, Asaba H, Kang MJ, et al. Low-density lipoprotein receptor-related protein 5 (LRP5) is essential for normal cholesterol metabolism and glucose-induced insulin secretion. Proc Natl Acad Sci U S A. 2003;100:229–34.
38. Babij P, Zhao W, Small C, et al. High bone mass in mice expressing a mutant LRP5 gene. J Bone Miner Res. 2003;18:960–74.
39. Westendorf JJ, Kahler RA, Schroeder TM. Wnt signaling in osteoblasts and bone diseases. Gene. 2004;341:19–39.
40. Holmen SL, Giambernardi TA, Zylstra CR, et al. Decreased BMD and limb deformities in mice carrying mutations in both Lrp5 and Lrp6. J Bone Miner Res. 2004;19:2033–40.
41. Little RD, Carulli JP, Del Mastro RG, et al. A mutation in the LDL receptor-related protein 5 gene results in the autosomal dominant high-bone-mass trait. Am J Hum Genet. 2002;70:11–9.
42. Akhter MP, Wells DJ, Short SJ, et al. Bone biomechanical properties in LRP5 mutant mice. Bone. 2004;35:162–9.

43. Johnson ML, Harnish K, Nusse R, Van Hul W. LRP5 and Wnt signaling: a union made for bone. J Bone Miner Res. 2004;19:1749–57.
44. Johnson ML, Summerfield DT. Parameters of LRP5 from a structural and molecular perspective. Crit Rev Eukaryot Gene Expr. 2005;15:229–42.
45. Mani A, Radhakrishnan J, Wang H, et al. LRP6 mutation in a family with early coronary disease and metabolic risk factors. Science (New York, NY). 2007;315:1278–82.
46. Kim DH, Inagaki Y, Suzuki T, et al. A new low density lipoprotein receptor related protein, LRP5, is expressed in hepatocytes and adrenal cortex, and recognizes apolipoprotein E. J Biochem. 1998;124:1072–6.
47. Magoori K, Kang MJ, Ito MR, et al. Severe hypercholesterolemia, impaired fat tolerance, and advanced atherosclerosis in mice lacking both low density lipoprotein receptor-related protein 5 and apolipoprotein E. J Biol Chem. 2003;278:11331–6.

Osteocardiology: Cellular Origins of Cardiac Calcification

7

Introduction

For years cardiac calcification was thought to be a passive degenerative phenomenon. The cellular origins of cardiac calcification are under intense investigation. Understanding of the cellular mechanisms of this valve lesion will present new cellular therapeutic options to slow disease progression. VICs are the most common cells in the valve and are distinct from other mesenchymal cell types in other organs. Native valve interstitial cells are the primary cell responsible for the development of valve calcification [1]. There are five phenotypes best represent the VIC family of cells because each of these phenotypes exhibits specific cellular functions essential in normal valve physiology and in pathological processes [1]. The phenotypes as embryonic progenitor endothelial/mesenchymal cells, quiescent VICs (qVICs), activated VICs (aVICs), progenitor VICs (pVICs), and osteoblastic VICs (obVICs) [1]. The categories of cell types identify the cellular phenotype, localization and stage of development of the cell [1]. Mesenchymal progenitor cells in the peripheral blood have been identified as circulating osteoblast-lineage cells that give rise to cells with characteristics of adipocytes, osteoclasts, fibroblasts, or osteoblasts [2–4]. Figure 7.1, describes the two major categories of cells, circulating osteogenic progenitor cell, and the native interstitial cell.

Previously, several studies have demonstrated that aortic valve calcification is associated with endochondral bone formation and an osteoblast bone-like phenotype. Bone and cartilage are major tissues in the vertebrate skeletal system, which is primarily composed of three cell types: osteoblasts, chondrocytes, and osteoclasts. In the developing embryo, osteoblast and chondrocytes both differentiate from common mesenchymal progenitors in situ, whereas osteoclasts are of hematopoietic origin and brought in later by invading blood vessels. Osteoblast differentiation and maturation lead to bone formation controlled by two distinct mechanisms: intramembranous and endochondral ossification, both starting from mesenchymal condensations as described in Chap. 5.

N.M. Rajamannan, *Osteocardiology*, DOI 10.1007/978-3-319-64994-8_7

Two osteoblast-specific transcripts have been identified: (1) Cbfa1 and (2) osteocalcin. The transcription factor Cbfa1 has all the attributes of a "master gene" differentiation factor for the osteoblast lineage and bone matrix gene expression. During embryonic development, Cbfa1 expression precedes osteoblast differentiation and is restricted to mesenchymal cells destined to become osteoblast. In addition to its critical role in osteoblast commitment and differentiation, Cbfa1 appears to control osteoblast activity, ie, the rate of bone formation by differentiated osteoblasts TIEG1 has recently been linked to Wnt signaling in the heart and bone [5, 6]. The regulatory mechanism of osteoblast differentiation from osteoblast progenitor cells as shown in described in Chap. 5, Fig. 7.1, into terminally differentiated cells is via a well-orchestrated and well-studied pathway that involves initial cellular proliferation events and then synthesis of bone matrix proteins, which requires the actions of specific paracrine/hormonal factors/BMP and the activation of the canonical Wnt pathway. In a previous study by Suda et al. [7], they have shown that these isolated COP cells can express BMP and can form bone in vivo. Confirming the hypothesis that these COP cells are capable of homing to sites of valve calcification and neovascularization and form bone. The studies to date indicate that the cellular origins of bone forming cells

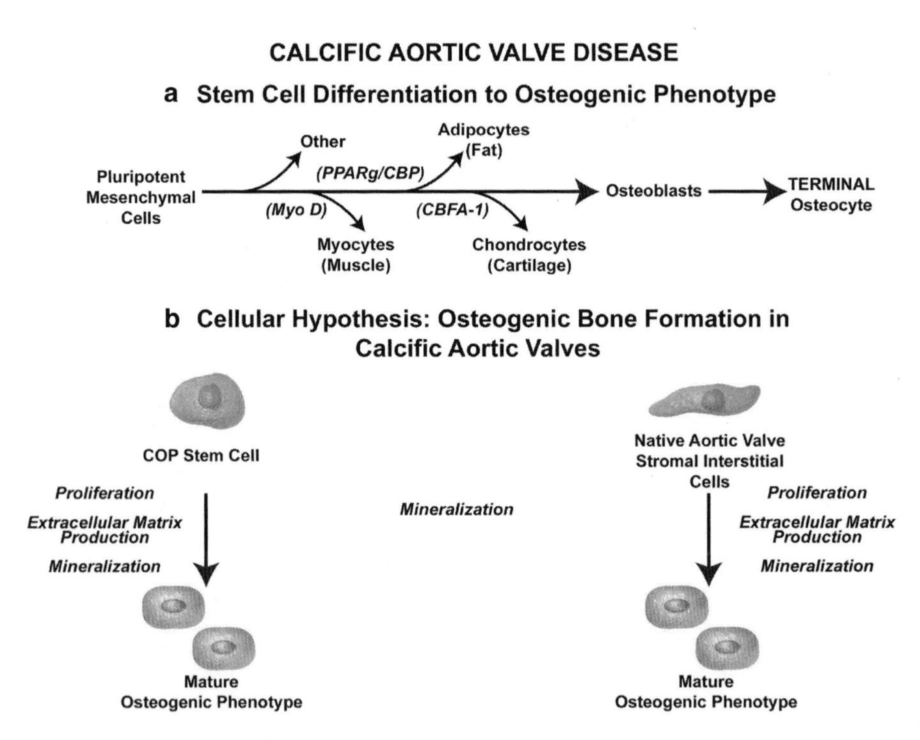

Fig. 7.1 Cellular origins of cardiac calcification demonstrates the calcific aortic valve disease: cellular origins of valves calcification and the mechanisms of osteogenesis in these cell types. (**a**) The potential cellular phenotypes for mesenchymal derived cells. (**b**) The two different cell origins for valve calcification: the COP cell and the native interstitial cell both contributing to the hypothesis of osteogenesis in calcific aortic valve disease. COP indicates circulating osteogenic precursor cells [8] (Permissions obtained for reproduction)

in the calcifying aortic valve have two distinct pathways as shown in Fig. 7.1 [8]. The cells can either be the COP cell capable of differentiating to bone at the site of calcification and disease [9], or the interstitial aortic valve cell that is capable of differentiating to bone in vivo, as described in the most recent National Heart and Lung and Blood Institute Working Group paper on calcific aortic valve disease.

Further evidence for the circulating stem cell was published in a study by Tanaka et al., which demonstrated using transplanted bone marrow cells composed of 17% of the population of calcifying cells in the native atherosclerotic valve in ApoE null mice [10]. The presence of variable depths of the COP cell in the calcific valve is consistent with the hypothesis that these cells can home to the diseased valve, but are not responsible for the entire bone formation process. The native interstitial cells also have the potential to differentiate to bone in situ and contribute to the calcifying cells in the native valve. The contribution of these two cell types toward the development of calcification in the aortic valve requires further ongoing investigation.

In the Mayo Clinic study, circulating mesenchymal osteoblast cells are isolated from peripheral blood via flow cytometry methods [4]. Bone specific antibodies were used to isolate osteopontin and alkaline phosphatase positive cells and then measured functional significance of the cells in young versus old male patients to determine the potential for osteogenesis [4]. The study investigators isolated osteoblast progenitor cells, and hypothesized the role of these cells in fracture healing and heterotopic bone formation in disease mechanisms [4]. In a study by Egan et al. [9], they identify for the first time in human calcifying aortic valves a population of circulating osteogenic precursor cells (COP) in calcified human aortic valves. Their finding of these CD45+ OCN+ COP cells in areas of calcification, and not in the unaffected calcified tissues provides another level of evidence that mesenchymal derived cell populations are responsible for the development of osteogenesis in the calcified aortic valve. Specifically, the study demonstrated that these cells were localized to areas of confirmed endochondral ossification and bone formation. Within the regions of interest there were areas of mature bone with the characteristic architecture including osteocytes and bone lining cells. However, within the limits of the study there was no consistent involvement of the valve leaflet layers as the areas of endochondral ossification were found to extend to variable depths. The conclusions from this study provides the first evidence in human calcifying aortic valve tissue that a novel cellular origin is found on the calcific aortic valve and that these COP cells play a role in the cellular mechanisms of osteogenesis.

Embryonic Implications of Cellular Origin of Precursor Cells

The low-density lipoprotein-related receptor 5 and 6 (Lrp5 and Lrp6) genes were cloned in 1998 based on their homology with the low-density lipoprotein receptor (LDLR) [11, 12–14]. Mutations in either LRP5 or LRP6 proteins have caused a number of disease processes in the field of bone [15, 16], and have been associated with cardiovascular disease [14, 17–19]. In the study by Borrell-Pages et al. [20] the authors confirm the novel finding that Lrp5 plays an atheroprotective role in the vascular aorta. In the wildtype (WT) versus the $Lrp5^{-/-}$ mice there are larger atheromatous lesions in the $Lrp5^{-/-}$ mice as compared to WT littermates, with an upregulation of the LDLR family members including VLDR, Lrp6 and Lrp2. The mechanism postulated by the authors implicates higher plasma cholesterol levels in the $Lrp5^{-/-}$ mice as compared to the WT littermates as the driving factor for the significant increase in atheroma in the thoracic aortas. The production of mice lacking Lrp5 revealed that Lrp5 deficiency led to increased plasma cholesterol levels in mice fed a high-fat diet, secondary to decreased hepatic clearance of chylomicron remnants and also marked impaired glucose tolerance [17]. Lrp6 also regulates bone, but has been found to have a low bone mass effect in patients in which a putative partial loss-of-function mutation in Lrp6 was identified to lead to early cardiovascular-related death associated with increased plasma LDL, triglycerides, hypertension, diabetes and osteoporosis [21]. This background studies are the foundation for the results in the novel study by Borrell-Pages et al., implicating the role of Lrp6, CLDR and Lrp2 in lipid metabolism and progression of aorta atherosclerosis [20].

Previous studies testing experimental hypercholesterolemia in mouse and rabbit models demonstrated an upregulation of Lrp5 receptor expression and activation of cell proliferation and extracellular matrix production critical in bone formation in the aortic valve *in vivo* and *ex vivo* [18, 19, 22]. Specificity for the role of Lrp5 in aortic valve calcification was tested in the previous study using a high cholesterol diet in the Lrp5 null mice, which demonstrated opposite results in the aortic valve: no evidence of atherosclerosis or valve calcification [23].

The Lrp5 pathway also regulates bone formation in different diseases of bone [15, 24]. The discovery that the Lrp5 receptor carries the gain of function [24] and loss of function [15] mutations in the development of bone diseases, resulted in a number of studies which have shown that activation of the canonical Wnt pathway is important in osteoblastogenesis [17, 25–27]. Three studies to date have confirmed the regulation of the Lrp5/Wnt pathway for cardiovascular calcification *in vivo* and *ex vivo* [18, 19, 22]. In this pathway, Wnt proteins bind to receptors composed of a frizzled protein and either of the low-density lipoprotein receptor-related proteins Lrp5 or Lrp6. Signaling via Disheveled and/or Axin then results in inactivation of a multiprotein complex including Axin, adenomatous polyposis coli (APC), and glycogen synthase kinase-3β that normally renders β-catenin unstable. By inhibiting this complex, Wnt signals lead to accumulation of β-catenin in the cytosol and its entry into the nucleus. Once in the nucleus, β-catenin binds to proteins of the T-cell factor/lymphoid enhancer factor-1 family and modulates the expression of several target genes which include Cyclin D, Cbfa1, and Sox9. Bone and cartilage are major tissues in the vertebrate skeletal system, which is primarily composed of three cell types: osteoblasts, chondrocytes, and osteoclasts. In the developing embryo, osteoblast and chondrocytes both differentiate from common mesenchymal progenitors *in situ*, whereas osteoclasts are of hematopoietic origin and brought in later by invading blood vessels. Osteoblast differentiation and maturation lead to bone formation controlled by two distinct mechanisms: intramembranous and endochondral ossification, both starting from mesenchymal condensations.

The role of lipid signaling of the Lrp5 receptor has been defined in experimental *in vitro* and *in vivo* lipid models of vascular atherosclerosis. Lrp5, which binds apoE-containing lipoproteins *in vitro*, is widely expressed in many tissues including hepatocytes, adrenal gland and pancreas [14]. The production of mice lacking Lrp5 revealed that Lrp5 deficiency led to increased plasma cholesterol levels in mice fed a high-fat diet, secondary to decreased hepatic clearance of chylomicron remnants, and also marked impaired glucose tolerance [17]. The Lrp5 deficient islets also demonstrated a reduction of intracellular ATP and Calcium in response to glucose, thereby decreasing glucose induced insulin secretion [17]. Furthermore, experimental hypercholesterolemia is associated with the increase in Lrp5 receptor expression and activation of cell proliferation and extracellular matrix production critical in bone formation [18]. These studies provide evidence that lipoprotein metabolism is regulated by the fifth family member of the LDL co-receptor family Lrp5 in these knockout mouse studies.

Embryonically, Wnt proteins bind to receptors composed of a frizzled protein and either of the low-density lipoprotein receptor-related proteins Lrp5 or Lrp6. In the developing embryo, osteoblast and chondrocytes, both differentiate from common mesenchymal progenitors, and neural crest cells. The role of Wnt and Lrp5 coreceptors in embryogenesis, have been the most detailed studies in the field to date [28]. Studies have demonstrated that the neural crest cells are specific to the aortic valve [28] and mesodermal cells are specific to the descending aorta [29]. Furthermore, a recent study in the proximal ascending aorta where neural crest cells reside, versus the descending aorta, indicate proximal aorta calcifies at an accelerated rate than the descending aorta in the presence of hyperphosphatemia [30].

Figure 7.2, demonstrates the role of hypercholesterolemia and the role of the Lrp5 receptor in the aortic valve and the aorta. The embryonic cell origin may

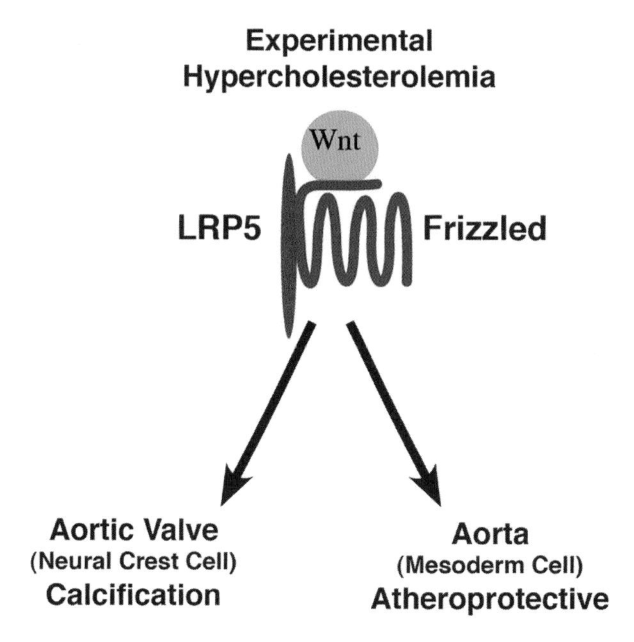

Fig. 7.2 Lrp5 is atheroprotective in the vascular aorta and osteogenic in the aortic valve in the presence of experimental hypercholesterolemia [37] (Permission obtained for reproduction)

provide another clue for why the hypercholesterolemic Lrp5 null mouse develops excessive atherosclerosis in the aorta that has mesodermal derived cells [20], but does not develop calcifying valve lesion in the aortic valve [23]. The novel finding by Borrell-Pages et al. [20], in this issue of Atherosclerosis that Lrp5 plays an atheroprotective role in the vascular aorta, is unique in the field and will help to further define the complex role of Lrp5 co-receptors in the field of embryogenesis, atherosclerosis, cell differentiation and bone biology.

Stem Cell Hypothesis in Bioprosthetic Valve Calcification

In an experimental hypercholesterolemic model, as the environment of hypercholesterolemia induces inflammation, mesenchymal stem cells mobilize and home to the implanted bioprosthetic valve. Atorvastatin attenuates these markers in the rabbit model. Bioprosthetic valve calcification is a unique model to study the role of stem cells in valve calcification as shown in Fig. 7.3 [31]. The finding of cKit mesenchymal expression, in atherosclerotic bioprosthetic valves provides a mechanism by which calcification can develop. [32] Tanaka et al. [10], have demonstrated in native aortic valves in hypercholesterolemic mice that 10% of cells are bone marrow derived cells within the atherosclerotic lesion. The rest of the cells are native

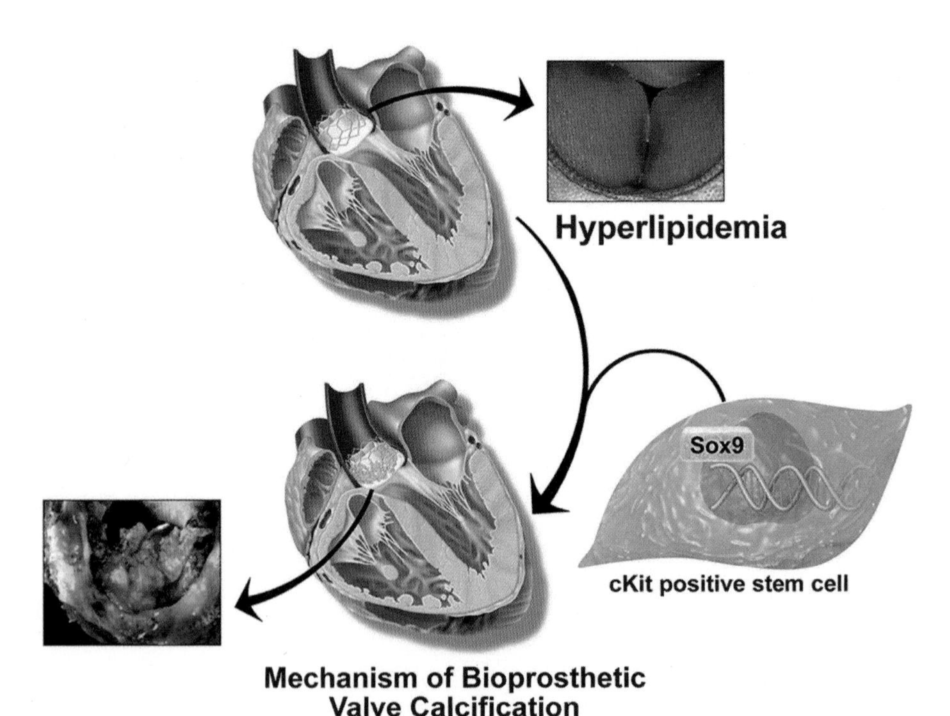

Fig. 7.3 Mechanisms of Heterotopic bone formation in heterotopic calcification in bioprosthetic heart valves [31]

myofibroblasts differentiating into an osteoblast phenotype as implicated in other studies of aortic valve disease [18, 33–36].

In Summary

In summary, this chapter provides incremental understanding into the role of cellular origins of cardiovascular calcification. In the future, these studies will provide the roadmap for targeted specific cellular phenotypes in slowing progression of calcification in the heart.

References

1. Liu AC, Joag VR, Gotlieb AI. The emerging role of valve interstitial cell phenotypes in regulating heart valve pathobiology. Am J Pathol. 2007;171:1407–18.
2. Zvaifler NJ, Marinova-Mutafchieva L, Adams G, et al. Mesenchymal precursor cells in the blood of normal individuals. Arthritis Res. 2000;2:477–88.
3. Kuznetsov SA, Mankani MH, Gronthos S, Satomura K, Bianco P, Robey PG. Circulating skeletal stem cells. J Cell Biol. 2001;153:1133–40.
4. Eghbali-Fatourechi GZ, Lamsam J, Fraser D, Nagel D, Riggs BL, Khosla S. Circulating osteoblast-lineage cells in humans. N Engl J Med. 2005;352:1959–66.
5. Subramaniam M, Cicek M, Pitel KS, et al. TIEG1 modulates beta-catenin sub-cellular localization and enhances Wnt signaling in bone. Nucleic Acids Res. 2017;45:5170–82.
6. Rajamannan NM. The role of TIEG1 in calcific aortic valve disease. Journal of Bone and Mineral Metabolism. 2017;29(9):S136.
7. Suda RK, Billings PC, Egan KP, et al. Circulating osteogenic precursor cells in heterotopic bone formation. Stem Cells. 2009;27:2209–19.
8. Rajamannan NM. Calcific aortic valve disease: cellular origins of valve calcification. Arterioscler Thromb Vasc Biol. 2011;31:2777–8.
9. Egan KP, Kim JH, Mohler ER 3rd, Pignolo RJ. Role for circulating osteogenic precursor cells in aortic valvular disease. Arterioscler Thromb Vasc Biol. 2011;31:2965–71.
10. Tanaka K, Sata M, Fukuda D, et al. Age-associated aortic stenosis in apolipoprotein E-deficient mice. J Am Coll Cardiol. 2005;46:134–41.
11. Dong Y, Lathrop W, Weaver D, et al. Molecular cloning and characterization of LR3, a novel LDL receptor family protein with mitogenic activity. Biochem Biophys Res Commun. 1998;251:784–90.
12. Brown SD, Twells RC, Hey PJ, et al. Isolation and characterization of LRP6, a novel member of the low density lipoprotein receptor gene family. Biochem Biophys Res Commun. 1998;248:879–88.
13. Hey PJ, Twells RC, Phillips MS, et al. Cloning of a novel member of the low-density lipoprotein receptor family. Gene. 1998;216:103–11.
14. Kim DH, Inagaki Y, Suzuki T, et al. A new low density lipoprotein receptor related protein, LRP5, is expressed in hepatocytes and adrenal cortex, and recognizes apolipoprotein E. J Biochem. 1998;124:1072–6.
15. Gong Y, Slee RB, Fukai N, et al. LDL receptor-related protein 5 (LRP5) affects bone accrual and eye development. Cell. 2001;107:513–23.
16. Little RD, Carulli JP, Del Mastro RG, et al. A mutation in the LDL receptor-related protein 5 gene results in the autosomal dominant high-bone-mass trait. Am J Hum Genet. 2002;70:11–9.
17. Fujino T, Asaba H, Kang MJ, et al. Low-density lipoprotein receptor-related protein 5 (LRP5) is essential for normal cholesterol metabolism and glucose-induced insulin secretion. Proc Natl Acad Sci U S A. 2003;100:229–34.

18. Rajamannan NM, Subramaniam M, Caira F, Stock SR, Spelsberg TC. Atorvastatin inhibits hypercholesterolemia-induced calcification in the aortic valves via the Lrp5 receptor pathway. Circulation. 2005;112:I229–34.
19. Caira FC, Stock SR, Gleason TG, et al. Human degenerative valve disease is associated with up-regulation of low-density lipoprotein receptor-related protein 5 receptor-mediated bone formation. J Am Coll Cardiol. 2006;47:1707–12.
20. Borrell-Pages M, Romero JC, Badimon L. Cholesterol modulates LRP5 expression in the vessel wall. Atheroslcrosis. 2014;235:363–70.
21. Mani A, Radhakrishnan J, Wang H, et al. LRP6 mutation in a family with early coronary disease and metabolic risk factors. Science (New York NY). 2007;315:1278–82.
22. Shao JS, Cheng SL, Pingsterhaus JM, Charlton-Kachigian N, Loewy AP, Towler DA. Msx2 promotes cardiovascular calcification by activating paracrine Wnt signals. J Clin Invest. 2005;115:1210–20.
23. Rajamannan NM. The role of Lrp5/6 in cardiac valve disease: experimental hypercholesterolemia in the ApoE−/− /Lrp5−/− mice. J Cell Biochem. 2011;112:2987–91.
24. Boyden LM, Mao J, Belsky J, et al. High bone density due to a mutation in LDL-receptor-related protein 5. N Engl J Med. 2002;346:1513–21.
25. Babij P, Zhao W, Small C, et al. High bone mass in mice expressing a mutant LRP5 gene. J Bone Miner Res. 2003;18:960–74.
26. Westendorf JJ, Kahler RA, Schroeder TM. Wnt signaling in osteoblasts and bone diseases. Gene. 2004;341:19–39.
27. Holmen SL, Giambernardi TA, Zylstra CR, et al. Decreased BMD and limb deformities in mice carrying mutations in both Lrp5 and Lrp6. J Bone Miner Res. 2004;19:2033–40.
28. Phillips HM, Mahendran P, Singh E, Anderson RH, Chaudhry B, Henderson DJ. Neural crest cells are required for correct positioning of the developing outflow cushions and pattern the arterial valve leaflets. Cardiovasc Res. 2013;99:452–60.
29. Zaniboni A, Bernardini C, Alessandri M, et al. Cells derived from porcine aorta tunica media show mesenchymal stromal-like cell properties in in vitro culture. Am J Physiol Cell Physiol. 2014;306:C322–33.
30. Leroux-Berger M, Queguiner I, Maciel TT, Ho A, Relaix F, Kempf H. Pathologic calcification of adult vascular smooth muscle cells differs on their crest or mesodermal embryonic origin. J Bone Miner Res. 2011;26:1543–53.
31. Rajamannan NM, et al. Atorvastatin attenuates bioprosthetic heart valve calcification in a rabbit model via stem cell mediated mechanism, Journal of American College of Cardiology. 2008;51(10):A277.
32. D'Ippolito G, Diabira S, Howard GA, Menei P, Roos BA, Schiller PC. Marrow-isolated adult multilineage inducible (MIAMI) cells, a unique population of postnatal young and old human cells with extensive expansion and differentiation potential. J Cell Sci. 2004;117:2971–81.
33. Rajamannan NM, Subramaniam M, Springett M, et al. Atorvastatin inhibits hypercholesterolemia-induced cellular proliferation and bone matrix production in the rabbit aortic valve. Circulation. 2002;105:2260–5.
34. Rajamannan NM, Subramaniam M, Stock SR, et al. Atorvastatin inhibits calcification and enhances nitric oxide synthase production in the hypercholesterolaemic aortic valve. Heart. 2005;91:806–10.
35. Aikawa E, Nahrendorf M, Sosnovik D, et al. Multimodality molecular imaging identifies proteolytic and osteogenic activities in early aortic valve disease. Circulation. 2007;115:377–86.
36. Weiss RM, Ohashi M, Miller JD, Young SG, Heistad DD. Calcific aortic valve stenosis in old hypercholesterolemic mice. Circulation. 2006;114:2065–9.
37. Rajamannan NM. Embryonic cell origin defines functional role of Lrp5. Atherosclerosis. 2014;236:196–7.

Osteocardiology: LDL-Density-Gene Theory

8

Introduction

Cardiovascular calcification is the end stage phenotype after years of subclinical atherosclerosis in the heart. For years, the mechanism for this calcification was thought to be due to a passive degenerative process. However, in the twenty-first century, the National Heart Lung and Blood Institute of the NIH, recognizes that calcific aortic valve disease, is an active biologic process [1]. The risk factors for the initiation event in calcific aortic valve disease, have been identified as traditional atherosclerotic risk factor well known to promote coronary artery disease (CAD) as well as CAVD. The calcification process involves normal valve interstitial cells differentiating via an osteogenic gene activation, which results in a calcified osteoblast like phenotype [2].

Familial Hypercholesterolemia Genetics and Cholesterol Metabolism

In 1985, Brown and Goldstein, received the Nobel Prize in Medicine for "their discoveries concerning the regulation of cholesterol metabolism." Their discovery defined the role of the LDL receptor in cholesterol metabolism and homozygous defect in patients is the mechanism for severe hypercholesterolemia [3]. In 2017, the clinical recognition and diagnosis of the homozygous hypercholesterolemia remains under diagnosed globally. The manifestations of chronic exposure to elevated cholesterol, is the initial event in abnormal oxidative stress, abnormal endothelial nitric oxide synthase function and eventual atherosclerosis and calcification in the cardiovascular system.

Since the initial discoveries in the field of FH genetics, several gene defects have been identified including a defective apoB100 component of LDL, known as familial defective apoB-100, clinically indistinguishable from heterozygous LDLR mutation [4]. The third identifiable genetic gain of function mutation identified

N.M. Rajamannan, *Osteocardiology*, DOI 10.1007/978-3-319-64994-8_8

affects the proprotein convertase subtilisin/kexin 9 (PCSK9) encoded by chromosome 1 has also been shown to initiate FH by negatively regulating the LDL receptor expression [5]. The genetics, diagnosis and clinical manifestations have been the subject of intense investigation and continued discoveries in the evolution of therapeutic interventions for this patient population.

Familial Hypercholesterolemia as a Genetic Model for CAC, CAVD, CAD

Familial hypercholesterolemia (FH) is a genetic disorder of lipoprotein metabolism resulting in elevated serum low-density lipoprotein (LDL) cholesterol levels leading to increased risk for premature cardiovascular calcification [6, 7]. The diagnosis of this condition is based on clinical features, family history, and elevated LDL-cholesterol levels aided more recently by genetic testing. As the atherosclerotic burden is dependent on the degree and duration of exposure to raised LDL-cholesterol levels, early diagnosis and initiation of treatment is critical. Statins are presently the mainstay in the management of these patients, although newer drugs, LDL apheresis, and other novel rapidly established therapies in recent clinical trials [8, 9], such as PCSK9 inhibition, will play a role in certain subsets of FH. Together these novel treatments have notably improved the prognosis of FH, especially that of the heterozygous patients. Despite these achievements, a majority of children fail to attain targeted lipid goals owing to persistent shortcomings in diagnosis, monitoring, and treatment [10].

FH and CAVD in MESA

In a recent study by ten Kate et al. [11], the objective was to investigate the prevalence, extent and risk modifiers of CAVD in heterozygous FH (he-FH) patients. Clinically, the heterozygous FH phenotype is encountered most often while the homozygous phenotype is has much worse clinical consequences, including premature CAD. The investigators sought to determine the prevalence of calcific aortic valve disease in patients with the more common finding of heterozygous FH, by performing and analysis which measured the amount of calcification burden via CT measurements and Agatston score of the coronary artery and aortic valve and to compare lipid levels to the amount of functional LDL receptor gene, and traditional cardiovascular risk factors in this patient population.

The results demonstrated the prevalence and Aortic valve calcium score (CAVD-score) were higher in the he-FH patients than in controls: 41% 51 (9–117);and 21% (3–49) ($p < 0.001$ and $p = 0.007$). Age, untreated maxLDL, CAC and diastolic blood pressure were independently associated with CAVD. LDLR-negative mutational he-FH was the strongest predictor of the CAVD-score (OR: 4.81; 95% CI: 2.22–1040; $p < 0.001$). Compared to controls, he-FH is associated with a high prevalence and a large extent of subclinical CAVD, especially in patients with LDLR-negative mutations.

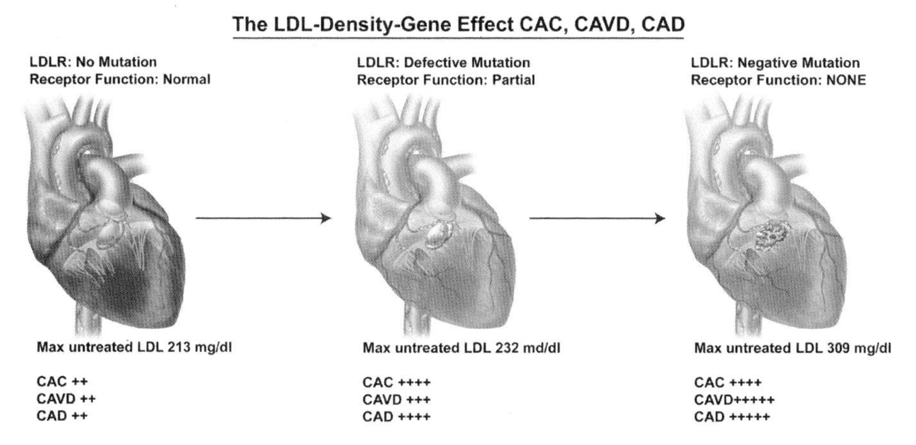

Fig. 8.1 LDL-Density-Gene Theory in cardiovascular calcification. The amount of functional LDLR correlates with the amount of calcification in the artery, valve and the aorta [6] (Permission obtained to reproduce)

Among the variables, LDLR-negative mutation carrier status was a strong predictor of the extent of the CAVD. Association between coronary and aortic valve calcification with the presence of CAC was associated with a higher prevalence of CAVD, both in he-FH and control patients. Compared to he-FH patients with LDLR-defective mutations, LDLR-negative mutational he-FH was associated with higher total cholesterol and untreated maxLDL, in addition he-FH patients with LDLR-negative mutations were younger started using statins at a younger age and used statins for a longer period of time. He-FH patients were LDLR-negative mutations had higher prevalence of CAVD as compared to LDLR-defective mutations and controls. The difference in CAVD prevalence between LDLR-defective mutational he-FH and the controls was also significant additionally CAVD scores increase faster with age in LDLR-negative he-FH than in the LDLR-defective he-FH. Figure 8.1, demonstrates the effect of the density of functional LDL receptors and proportional increases in cholesterol with the degree of aortic valve calcifications, coronary artery calcification and calcific aortic disease as described by the Montreal group and the identification of a gene dosage effect in homozygote versus heterozygote FH patients [12].

Familial Hypercholesterolemia and Calcific Aortic Disease

In a recent study by Kindi et al. [13], the investigators sought to determine the rate of progression of aorta calcification in patients with HeFH. Sixteen HeFH patients, all with the null LDLR DEL15Kb mutation were studied using thoraco-abdominal CT scans and quantification scores. Patients were scanned at baseline and rescanned an average of 8.2 ± 0.8 years after the first scan. Mean LDL-C was 2.53 mmol/L on medical therapies. Aortic calcification increased in all patients in an exponential fashion with respect to age. Age was the strongest correlate of AoCa score. Investigators studied only patients heterozygous for FH and analyzed the data using age statistical analysis scores. The effect of age demonstrated a fivefold increase in the progression in the HeFH patient population with an exponential increase only correlating with age as a risk factor. Despite the known defect in lipid metabolism in this subset of FH patients, traditional risk factor of lipids did not play a role in the progression of the calcification process in this patient population. Furthermore, the effect of lipid lowering of 65% reduction in the LDL-C from the baseline values, medical therapy did not attenuate the process. The results of this study indicate that patients who have the diagnosis of HeFH, and are delayed to the time of diagnosis, the calcification process will progress even on optimal medical therapy to lower lipids.

Phenotypic Expression of Calcification in the Heart: The Bernoulli Equation

Results from this study demonstrate the first correlation of the role of LDL, genetic contribution of LDL in terms of function and phenotypic expression of calcification in the valve and in the coronary arteries. Fluid hemodynamics in the heart is dependent on multiple factors as derived by the Bernoulli's equation for fluid flow [14]. Bernoulli described flow through a column is directly proportional to the change in pressure across the column and indirectly proportional to the resistance. The formula for flow through the heart, is similar to Ohm's law for electricity as shown in Eq. 8.1.

$$\mathrm{Q} = \frac{\Delta P}{\mathrm{R}} \tag{8.1}$$

The entire formula for resistance for steady state flow through a circular tube, is shown in Eq. 8.2, where η = viscosity, r = radius of the tube.

$$\mathrm{R} = \frac{8\eta L}{\pi \mathrm{r}^4} \tag{8.2}$$

Equations 8.1 and 8.2 can be combined to give the flow rate through a circular tube in terms of a pressure drop which is described as Poiseuille's law:

$$\mathrm{Q} = \frac{\pi \mathrm{r}^4}{8\eta L} \Delta P \tag{8.3}$$

The differences in the rate of fluid flow are dependent on the radius of the anatomic structure, which is inversely proportional to the resistance. In addition, it is important to note the inverse r^4 dependence of the resistance to fluid flow. If the radius of the tube is halved, the pressure drop for a given flow rate and viscosity is increased by a factor of 16. Since the flow rate is then proportional to the fourth power of the radius. The size of the radius becomes important as blood flows through the heart.

Summary

The LDL-Density-Radius Theory [15] and the LDL-Density-Pressure Theory [16] hypothesize the role of lipids in the differentiation of valve myofibroblasts into osteoblast like cells responsible for the calcifying phenotype. The LDL-Density-Gene Theory correlates the role of functional gene receptor in the development of cardiovascular calcification.

References

1. Rajamannan NM, Evans FJ, Aikawa E, Grande-Allen KJ, Demer LL, Heistad DD, Simmons CA, Masters KS, Mathieu P, O'Brien KD, Schoen FJ, Towler DA, Yoganathan AP, Otto CM. Calcific aortic valve disease: not simply a degenerative process: a review and agenda for research from the National Heart and Lung and Blood Institute Aortic Stenosis Working Group. Executive summary: calcific aortic valve disease-2011 update. Circulation. 2011;124:1783–91.
2. Rajamannan NM, Subramaniam M, Rickard D, Stock SR, Donovan J, Springett M, Orszulak T, Fullerton DA, Tajik AJ, Bonow RO, Spelsberg T. Human aortic valve calcification is associated with an osteoblast phenotype. Circulation. 2003;107:2181–4.
3. Goldstein JL, Brown MS. Binding and degradation of low density lipoproteins by cultured human fibroblasts. Comparison of cells from a normal subject and from a patient with homozygous familial hypercholesterolemia. J Biol Chem. 1974;249:5153–62.
4. Rader DJ, Cohen J, Hobbs HH. Monogenic hypercholesterolemia: new insights in pathogenesis and treatment. J Clin Invest. 2003;111:1795–803.
5. Cohen JC, Boerwinkle E, Mosley TH Jr, Hobbs HH. Sequence variations in PCSK9, low LDL, and protection against coronary heart disease. N Engl J Med. 2006;354:1264–72.
6. Rajamannan NM. Calcific aortic valve disease in familial hypercholesterolemia: the LDL-density-gene effect. J Am Coll Cardiol. 2015;66:2696–8.
7. Rajamannan NM, Spelsberg TC, Moura LM. Mitral valve disease in a patient with familial hypercholesterolemia. Rev Port Cardiol. 2010;29:841–2.
8. Castilla-Guerra L, Fernandez-Moreno MC. PCSK9 inhibitors: a new era in stroke prevention? Eur J Intern Med. 2017;37:e44.
9. Ito MK, Santos RD. PCSK9 inhibition with monoclonal antibodies: modern management of hypercholesterolemia. J Clin Pharmacol. 2017;57:7–32.
10. Varghese MJ. Familial hypercholesterolemia: a review. Ann Pediatr Cardiol. 2014;7:107–17.
11. ten Kate G, Bos S, Dedic A, Neefjes L, Kurata A, Langendonk J. Increased aortic valve calcification in familial hypercholesterolemia: prevalence, extent and associated risk factors in a case-control study. J Am Coll Cardiol. 2015;66:2687–95.
12. Awan Z, Alrasadi K, Francis GA, Hegele RA, McPherson R, Frohlich J, Valenti D, de Varennes B, Marcil M, Gagne C, Genest J, Couture P. Vascular calcifications in homozygote familial hypercholesterolemia. Arterioscler Thromb Vasc Biol. 2008;28:777–85.
13. Kindi MA, Belanger AM, Sayegh K, Senouci S, Aljenedil S, Sivakumaran L, Ruel I, Rasadi KA, Waili KA, Awan Z, Valenti D, Genest J. Aortic calcification progression in heterozygote familial hypercholesterolemia. Can J Cardiol. 2017;33:658–65.
14. Bernoulli D. Hydrodynamica sive de viribus et motibus fluidorum commentarrii. Strasbourg: Argentoratum; 1738. p. St31.
15. Rajamannan NM. Mechanisms of aortic valve calcification: the LDL-density-radius theory: a translation from cell signaling to physiology. Am J Physiol Heart Circ Physiol. 2010;298:H5–15.
16. Rajamannan NM. The role of Lrp5/6 in cardiac valve disease: LDL-density-pressure theory. J Cell Biochem. 2011;112:2222–9.

Osteocardiology: The LDL-Density-Mechanostat Theory

9

Introduction

Left-sided valvular heart disease is the number one indication for valve interventions in the world. The cellular mechanisms of the most common left sided valve lesions from calcific aortic valve disease and to myxomatous mitral valve disease are under intense investigation. The signaling pathways described in left sided valvular heart disease have increased our understanding of this active biology. An understanding of these pathways may provide a basis towards understanding the disease process, the disease progression and the design of future clinical trials towards slowing the progression of disease. This unique hypothesis may help to define future clinical trials in this field of osteocardiology.

Left sided valve lesions are the most common indication for heart valve disease in the world. For years, this disease process was thought to be due to a passive degeneration of the heart valve. However, the cellular mechanisms have been under intense investigation in the twenty-first century. Recent studies have demonstrated that left-sided valve lesions develop aortic valve calcification and mitral valve myxomatous changes secondary to oxidative stress. Experimental models have demonstrated the phenotype associated with this hypothesis, with calcification in the aortic valve and cartilage formation in the mitral position. This chapter will review the common signaling and hemodynamic mechanisms, which will define the phenotypic expression of bone formation in the heart.

Osteocardiology Risk Factors

Over the past century, risk factors in the development of cardiovascular disease have become the central focus of experimental models translating to clinical therapies to treat the progression of cardiovascular disease including: vascular atherosclerosis [1], calcific aortic valve disease [2] and myxomatous mitral valve disease [3].

N.M. Rajamannan, *Osteocardiology*, DOI 10.1007/978-3-319-64994-8_9

Therapeutic options for lipid related cardiovascular disease include statin therapy [4], PSK9 antibody therapy [5], and other lipid lowering options.

Cellular Mechanisms of Left-Sided Valve Disease: The Osteogenic Phenotype

The hallmark of aortic valve stenosis is calcification, which for years was thought to be due to a passive phenomenon, but currently is defined as a bone formation process. The hallmark of myxomatous mitral valve disease was thought to be due to a passive mechanism, but currently is defined as a cartilage formation process. The traditional cardiovascular risk factors [1, 6–21] are important in the development of the final common pathway: the osteogenic phenotype [22, 23]. These studies show that there are evolving mechanisms for aortic valve calcification and myxomatous mitral valve disease, include cardiovascular risk factors and genetics to active specific cell signaling pathways in this disease process.

Experimental Mouse Model of Aortic Valve Calcification and Femur Osteoporosis

Chronic experimental hypercholesterolemia models develop aortic valve atherosclerosis and eventual calcification secondary to myofibroblast differentiation, which provides further direction for the understanding of the initiating events in the disease development [2, 24–35]. To test the hypothesis if LDLR$^{-/-}$ mice were treated a cholesterol diet. The valves developed atherosclerosis in the valve and osteoporosis in the bone via cholesterol activation of the Lrp5 pathway. LDLR$^{-/-}$ mice were given a cholesterol diet versus cholesterol and atorvastatin [36]. Atorvastatin modifies this process [36]. In the presence of cardiovascular risk factors as calcification develops in the aortic valve and the mineralization decreases in the femur.

Experimental Model of Mitral Valve Cartilage Formation and Regurgitation

Chronic experimental hypercholesterolemia models develop mitral valve atherosclerosis and eventual chondrocyte formation, which provide further direction for the understanding of the initiating events in the disease development [3, 37, 38]. To test the hypothesis if LDLR$^{-/-}$ rabbits would develop mitral regurgitation, the rabbits were given a cholesterol diet [39]. The mitral valves develop atherosclerosis in the mitral valve and regurgitation via cholesterol activation of the Lrp5 pathway, which is attenuated with atorvastatin.

Lrp5 Mechanostat Theory: The Role of Lrp5 in the Bone and Heart

The Lrp5 pathway regulates bone formation in different diseases of bone [40, 41]. The discovery of the Lrp5 receptor in the gain of function [41] and loss of function [40] mutations in the development of bone diseases, resulted in a number of studies which have shown that activation of the canonical Wnt pathway is important in osteoblastogenesis [42–45]. Several studies to date have confirmed the regulation of the Lrp5/Wnt pathway for cardiovascular calcification *in vivo* and *ex vivo* [34, 38, 46–52]. The concept of the mechanostat theory in the role of the Lrp5 bone receptor in the development of valve disease and bone disease is the paradox in the understanding of lipids in osteoporosis and valve disease. Global deletion of Lrp5 in mice results in significantly lower bone mineral density. Since osteocytes are proposed to act as a mechanosensor in the bone, investigators [53] addressed a question whether a conditional loss-of-function mutation of the Lrp5 receptor specific to osteocytes (Dmp1-Cre;Lrp5(f/f)) would alter responses to bone loading. In this experimental study, investigators tested loading to the right ulna for 3 min at a peak force of 2.65 N for 3 consecutive days, and the contralateral ulna was used as a non-loaded control. Young's modulus was determined using a section of the femur a common approach to measure load in bone biologic studies. The results showed that compared to age-matched littermate controls, mice lacking Lrp5 in osteocytes exhibited smaller skeletal size with reduced bone mineral density and content. Further proving the role of force on the Lrp5 receptor in the development of bone formation. Moreover, the results support the concept that the loss-of-function mutation of Lrp5 causing reduction of mechano-responsiveness and reduces bone mass and Young's modulus, and further supports the role of force in the biologic regulation of the Lrp5 receptor.

Phenotypic Expression of Calcification in the Heart: The Bernoulli Equation

A recent study [54] is the first to correlate the role of LDL and the effect of the LDL receptor genetic contribution in terms of phenotypic expression of calcification in the valve and in the coronary arteries. The LDL-Density theories [50, 55, 56] provide a hemodynamic explanation for why abnormal calcification develops secondary to high LDL density concentration up-regulating osteogenesis. The modulation of the hemodynamics by the variation of the vasculature radius is hypothesized to be responsible for the variable phenotype expression along the vasculature. This hypothesis may explain the reason why the aortic valve (high-pressure valve) and mitral valve (low-pressure valve) follow fundamentally different disease pathways namely, calcification and chondrocyte formation, respectively [38].

Hemodynamics in the heart is dependent on multiple factors, as derived by the Bernoulli's equation for fluid flow [57]. Bernoulli described that the flow through a cylinder is directly proportional to the change in pressure across the cylinder and indirectly proportional to the resistance as shown in Eq. 9.1.

$$\mathrm{Q} = \frac{\Delta P}{\mathrm{R}} \tag{9.1}$$

The entire formula for resistance for steady state flow through a circular tube, is shown in Eq. 9.2, where η = viscosity, r = radius of the tube.

$$\mathrm{R} = \frac{8\eta L}{\pi \mathrm{r}^4} \tag{9.2}$$

Equations 9.1 and 9.2 can be combined to give the flow rate through a circular tube in terms of a pressure drop, which is described as Poiseuille's law:

$$\mathrm{Q} = \frac{\pi \mathrm{r}^4}{8\eta L} \Delta P \tag{9.3}$$

The differences in the rate of fluid flow are dependent on the radius of the anatomic structure, which is inversely proportional to the resistance. In addition, it is important to note the inverse r^4 dependence of the resistance to fluid flow. If the radius of the tube is halved, the pressure drop for a given flow rate and viscosity is increased by a factor of 16. The LDL-Density-Radius Theory [55] and the LDL-Density-Pressure Theory [50] provide the molecular hypothesis for the role of lipids in the differentiation of valve myofibroblasts into osteoblast-like cells responsible for the calcifying phenotype. These theories also provide the fundamental basis for the faster rate of calcification observed in the coronary artery than in the aortic valve, based on the difference of radius at these two anatomic locations in the heart [58].

Role of Oxidative Stress in Left Sided Heart Valve Disease

Jiang et al. [59], published a careful examination of a surgical database of patients who underwent cardiac valve replacement. In the study, 400 patients with valvular heart disease, who received valve replacement, were found to have primarily left sided valve lesions. The investigators identified 77 of the patients in the entire surgical group with pathologic diagnosis of calcification, and the other 323 patients pathologic diagnosis without calcification.

In the calcification group, the majority of the valve replacements were aortic valve as compared to the mitral valve (25.9% vs 12.7%). The pathologic diagnosis for the patients in both the calcification and non-calcification group was rheumatic valve disease, with a higher incidence of congenital, degenerative and undefined valvular heart disease in the non-calcification group. In the calcified group the majority of the patients required aortic valve replacement, as compared to the non-calcified group, which the majority of patients with valve replacement were rheumatic mitral valve disease. The findings of increased calcification in the aortic valve versus mitral valve rheumatic patients, corresponds with a gradient of the calcified lesions developing on the left side. The final calcified phenotype develops at a greater amount in the aortic valve, which hemodynamically sustains higher pressures versus the lower pressure mitral valve in the cardiac chambers.

In terms of serum markers, the serum levels of alanine aminotransferase, aspartate aminotransferase, total bilirubin, total cholesterol, triglycerides, albumin, and serum creatinine were comparable. The serum BUN level was higher than in the control group. Age, calcium Gamma-glutamyl transferase (GGT), Retinol Binding Protein (RBP), and UA levels was chosen as factors for valve calcification and stratified for logistic regression analysis. Older age was tested as a risk factor age greater than 60 years Serum GGT and RBP were associated with increased valve calcification. Serum GGT between 30 and 46 IU/L showed a significant odds ratio for valve calcification (OR 2.771, 95% CI 1.333–5.759, $p = 0.0063$). The relative risk ratio for RBP for valve calcification was concentration-dependent but only serum RBP levels larger than 70 mg/mL showed a statistically significant difference (OR 4.110, 95% CI 1.452–11.637, $p = 0.0078$).

Interestingly serum calcium levels appeared to have a protective association with valve calcification. Calcium levels between 2.3 and 2.4 mmol/L showed the significantly protective effects on valve calcification (OR 0.270, 95% CI 0.082–0.889, $p = 0.313$). Serum ALP levels was negatively association with valve calcification. Finally, UA had no association with the calcification in the valves. Investigators have utilized experimental models to understanding the cellular mechanisms of calcific aortic valve disease [60]. These models test the role of oxidative stress in the development of aortic valve disease similar to those of vascular atherosclerosis [51]. The end-stage phenotype of this disease is an osteoblast calcifying phenotype. Candidate gene studies identified the role of Lrp5/6 co-receptors and Wnt signaling in the regulation of osteoblastogenesis in the bone, metabolic risk factors and lipid metabolism. Recent studies have demonstrated that Lrp5/6 plays a role in the development of valvular heart disease [34, 48, 50].

Stewart et al. [1, 19], described the risk factors for calcific aortic valve disease identified in the Cardiovascular Health Study. The investigators demonstrated that the clinical risk factors important for the development of atherosclerosis are also the independent risk factors for aortic valve stenosis including age, male gender, height (inverse relationship), history of hypertension, smoking and elevated serum levels of lipoprotein(a) and LDL levels [1]. Lipids and other cardiovascular risk factors induce oxidative stress [34] in the aortic valve endothelium similar to vascular endothelium [61] which in turn activates the secretion of cytokines and growth factors important in cell signaling. The early atherosclerotic and abnormal oxidative stress environment also plays a role in the activation of the calcification process in the myofibroblast cell via the Lrp5 receptor.

In the development of calcification in the left sided cardiac valves, a cascade of events similar to atherosclerosis must develop. In the presence of oxidative stress the aortic valve endothelium is activated and abnormal oxidation states develop.

The myofibroblast cells then begin to proliferate and synthesize extracellular bone matrix proteins with the upregulation of the Wnt/Lrp5 activation [34, 46]. These proteins overtime mineralize and calcify and a calcified aortic valve develops. The LDL-Density-Pressure theory provides the hemodynamic and embryologic basis for the role of Lrp5/6 signaling in calcific aortic valve disease.

LDL-Density-Pressure Theory: The Role of Oxidative Stress in the Left Sided Valve Lesions

This study confirms the hypothesis proposed in the LDL-Density-Pressure Theory, [50] that oxidative stress promotes the development of valve calcification in the aortic valve greater than the mitral valve. The effect of pressure in the development of calcification is dependent on two axioms, biology and hemodynamics. The clinical data demonstrates that oxidative stress activates the bone differentiation [34] in the valve. The initiating event of oxidative stress affects the left sided heart valves, as shown in Fig. 9.1. In the presence of oxidative stress only manifests disease in the left sided valves where the pressure in the heart is Overtime, the leaflets fuse which to occlude the aortic valve greater than the mitral valve. Furthermore, in the presence of elevated serum biomarkers of oxidative stress as defined in this study GGT and RBP, the development of calcification occurs in the aortic valve greater than the mitral valve. Furthermore, in this study, the rheumatic and degenerative valve

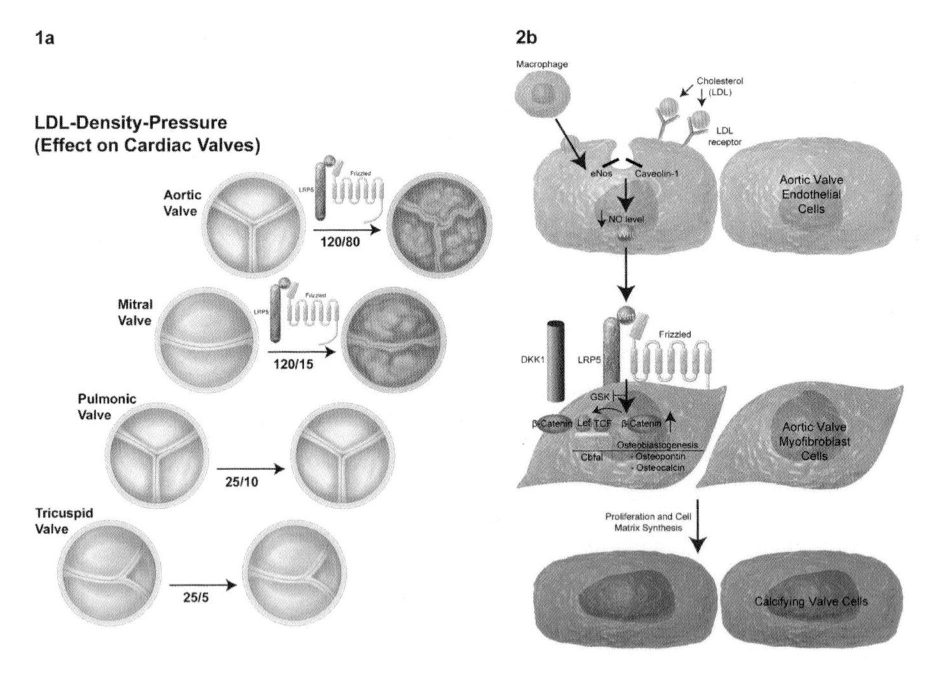

Fig. 9.1 The role of lipids and force on the regulation of Lrp5 in the development of calcification on the left side of the heart

population demonstrated a propensity of aortic valve calcification greater than mitral valve calcification in the rheumatic patients. Finally, these patients did not develop right-sided valve lesions.

For the past 40 years, catheter hemodynamics, echocardiography and timing of surgery have evolved as the diagnostic and therapeutic approach for calcific aortic stenosis. In the past decade, with the advent of experimental models and genetic testing, recognition that the aortic valve has an active cellular biology which incorporates three main processes for the development of calcific aortic stenosis, which include traditional cardiovascular risk factors, genetics and cellular signaling pathways to differentiate the valve into the osteoblast phenotype. The future management of this disease process will include the understanding of these different mechanisms for future medical therapy of this disease. If the physician can measure the novel risk factors such as GGT and RBP, in patients who present with an aortic valve murmur then targeting these risk factors may slow progression. The stethoscope can become an inexpensive screening tool for this pathologic process and possible preclinical atherosclerosis. If there are no identifiable risk factors then genetic considerations may play a role. Progress in this field, will make a difference for the future delay in the timing of surgery for these patients in the future.

Summary

The understanding of these signaling pathway and atherosclerotic risk factors and the role of blood pressure, is shown in Fig. 9.1, which outlines the interactions of the interrelated concept of traditional cardiovascular risk factors, and identified cellular targets for the treatment of left sided valve disease with increase activation of Lrp5 receptor complex in the presence of lipids with an increase in activation with higher pressures. Fig. 9.1, demonstrates the normal valve endothelial cells is located at the top of the figure. In the presence of lipids, the valve endothelium is activated and abnormal oxidation states develop. The myofibroblast cells then begin to proliferate and synthesize extracellular bone matrix proteins with the upregulation of the Wnt/Lrp5 activation [34, 46, 47]. These proteins overtime mineralize and calcify and a calcified aortic valve develops as shown at the bottom of the figure. The changes in pressure is a gradient in the heart, as the pressure increases in the left sided of the heart the force on the Lrp5 receptor increases to upregulate over time more calcification in the aortic valve and cartilage in the mitral valve and no evidence of disease in the right sided valves due to low pressures.

References

1. Stewart BF, Siscovick D, Lind BK, et al. Clinical factors associated with calcific aortic valve disease. Cardiovascular Health Study. J Am Coll Cardiol. 1997;29:630–4.
2. Rajamannan NM, Sangiorgi G, Springett M, et al. Experimental hypercholesterolemia induces apoptosis in the aortic valve. J Heart Valve Dis. 2001;10:371–4.

3. Rajamannan NM, Spelsberg TC, Moura LM. Mitral valve disease in a patient with familial hypercholesterolemia. Rev Port Cardiol. 2010;29:841–2.
4. Pedersen TR. Secondary prevention of coronary heart disease with lipid lowering drugs. Evaluating outcome eliminates the controversy. Der Internist (Berl). 1995;36:1174–8.
5. Kolodziejczak M, Navarese EP. Role of PCSK9 antibodies in cardiovascular disease: critical considerations of mortality and neurocognitive findings from the current literature. Atherosclerosis. 2016;247:189–92.
6. Deutscher S, Rockette HE, Krishnaswami V. Diabetes and hypercholesterolemia among patients with calcific aortic stenosis. J Chronic Dis. 1984;37:407–15.
7. Hoagland PM, Cook EF, Flatley M, Walker C, Goldman L. Case-control analysis of risk factors for presence of aortic stenosis in adults (age 50 years or older). Am J Cardiol. 1985;55:744–7.
8. Aronow WS, Ahn C, Kronzon I, Goldman ME. Association of coronary risk factors and use of statins with progression of mild valvular aortic stenosis in older persons. Am J Cardiol. 2001;88:693–5.
9. Aronow WS, Schwartz KS, Koenigsberg M. Correlation of serum lipids, calcium, and phosphorus, diabetes mellitus and history of systemic hypertension with presence or absence of calcified or thickened aortic cusps or root in elderly patients. Am J Cardiol. 1987;59:998–9.
10. Mohler ER, Sheridan MJ, Nichols R, Harvey WP, Waller BF. Development and progression of aortic valve stenosis: atherosclerosis risk factors—a causal relationship? A clinical morphologic study. Clin Cardiol. 1991;14:995–9.
11. Lindroos M, Kupari M, Valvanne J, Strandberg T, Heikkila J, Tilvis R. Factors associated with calcific aortic valve degeneration in the elderly. Eur Heart J. 1994;15:865–70.
12. Boon A, Cheriex E, Lodder J, Kessels F. Cardiac valve calcification: characteristics of patients with calcification of the mitral annulus or aortic valve. Heart. 1997;78:472–4.
13. Chui MC, Newby DE, Panarelli M, Bloomfield P, Boon NA. Association between calcific aortic stenosis and hypercholesterolemia: is there a need for a randomized controlled trial of cholesterol-lowering therapy? Clin Cardiol. 2001;24:52–5.
14. Wilmshurst PT, Stevenson RN, Griffiths H, Lord JR. A case-control investigation of the relation between hyperlipidaemia and calcific aortic valve stenosis. Heart. 1997;78:475–9.
15. Chan KL, Ghani M, Woodend K, Burwash IG. Case-controlled study to assess risk factors for aortic stenosis in congenitally bicuspid aortic valve. Am J Cardiol. 2001;88:690–3.
16. Briand M, Lemieux I, Dumesnil JG, et al. Metabolic syndrome negatively influences disease progression and prognosis in aortic stenosis. J Am Coll Cardiol. 2006;47:2229–36.
17. Palta S, Pai AM, Gill KS, Pai RG. New insights into the progression of aortic stenosis: implications for secondary prevention. Circulation. 2000;101:2497–502.
18. Peltier M, Trojette F, Sarano ME, Grigioni F, Slama MA, Tribouilloy CM. Relation between cardiovascular risk factors and nonrheumatic severe calcific aortic stenosis among patients with a three-cuspid aortic valve. Am J Cardiol. 2003;91:97–9.
19. Otto CM, Lind BK, Kitzman DW, Gersh BJ, Siscovick DS. Association of aortic-valve sclerosis with cardiovascular mortality and morbidity in the elderly [comment]. N Engl J Med. 1999;341:142–7.
20. Faggiano P, Antonini-Canterin F, Baldessin F, Lorusso R, D'Aloia A, Cas LD. Epidemiology and cardiovascular risk factors of aortic stenosis. Cardiovasc Ultrasound. 2006;4:27.
21. Pohle K, Maffert R, Ropers D, et al. Progression of aortic valve calcification: association with coronary atherosclerosis and cardiovascular risk factors. Circulation. 2001;104:1927–32.
22. Mohler ER 3rd, Gannon F, Reynolds C, Zimmerman R, Keane MG, Kaplan FS. Bone formation and inflammation in cardiac valves. Circulation. 2001;103:1522–8.
23. Rajamannan NM, Subramaniam M, Rickard D, et al. Human aortic valve calcification is associated with an osteoblast phenotype. Circulation. 2003;107:2181–4.
24. Rajamannan NM, Subramaniam M, Springett M, et al. Atorvastatin inhibits hypercholesterolemia-induced cellular proliferation and bone matrix production in the rabbit aortic valve. Circulation. 2002;105:2260–5.
25. Antonini-Canterin F, Hirsu M, Popescu BA, et al. Stage-related effect of statin treatment on the progression of aortic valve sclerosis and stenosis. Am J Cardiol. 2008;102:738–42.

26. Drolet MC, Arsenault M, Couet J. Experimental aortic valve stenosis in rabbits. J Am Coll Cardiol. 2003;41:1211–7.
27. Weiss RM, Ohashi M, Miller JD, Young SG, Heistad DD. Calcific aortic valve stenosis in old hypercholesterolemic mice. Circulation. 2006;114:2065–9.
28. Aikawa E, Nahrendorf M, Sosnovik D, et al. Multimodality molecular imaging identifies proteolytic and osteogenic activities in early aortic valve disease. Circulation. 2007;115:377–86.
29. Drolet MC, Roussel E, Deshaies Y, Couet J, Arsenault M. A high fat/high carbohydrate diet induces aortic valve disease in C57BL/6J mice. J Am Coll Cardiol. 2006;47:850–5.
30. Sarphie TG. Interactions of IgG and beta-VLDL with aortic valve endothelium from hypercholesterolemic rabbits. Atheroscelerosis. 1987;68:199–212.
31. Sarphie TG. A cytochemical study of the surface properties of aortic and mitral valve endothelium from hypercholesterolemic rabbits. Exp Mol Pathol. 1986;44:281–96.
32. Sarphie TG. Anionic surface properties of aortic and mitral valve endothelium from New Zealand white rabbits. Am J Anat. 1985;174:145–60.
33. Cimini M, Boughner DR, Ronald JA, Aldington L, Rogers KA. Development of aortic valve sclerosis in a rabbit model of atherosclerosis: an immunohistochemical and histological study. J Heart Valve Dis. 2005;14:365–75.
34. Rajamannan NM, Subramaniam M, Caira F, Stock SR, Spelsberg TC. Atorvastatin inhibits hypercholesterolemia-induced calcification in the aortic valves via the Lrp5 receptor pathway. Circulation. 2005;112:I229–34.
35. Miller JD, Chu Y, Brooks RM, Richenbacher WE, Pena-Silva R, Heistad DD. Dysregulation of antioxidant mechanisms contributes to increased oxidative stress in calcific aortic valvular stenosis in humans. J Am Coll Cardiol. 2008;52:843–50.
36. Rajamannan NM. Atorvastatin attenuates bone loss and aortic valve atheroma in LDLR mice. Cardiology. 2015;132:11–5.
37. Makkena B, Salti H, Subramaniam M, et al. Atorvastatin decreases cellular proliferation and bone matrix expression in the hypercholesterolemic mitral valve. J Am Coll Cardiol. 2005;45:631–3.
38. Rajamannan NM. Myxomatous mitral valve disease bench to bedside: LDL-density-pressure regulates Lrp5. Expert Rev Cardiovasc Ther. 2014;12:383–92.
39. Rajamannan N, Park J, Antonini-Canterin F. Development of an experimental model of mitral valve regurgitation via hypertrophic chondrocytes. Mol Biol Valv Heart Dis. 2014;1:35–41.
40. Gong Y, Slee RB, Fukai N, et al. LDL receptor-related protein 5 (LRP5) affects bone accrual and eye development. Cell. 2001;107:513–23.
41. Boyden LM, Mao J, Belsky J, et al. High bone density due to a mutation in LDL-receptor-related protein 5. N Engl J Med. 2002;346:1513–21.
42. Fujino T, Asaba H, Kang MJ, et al. Low-density lipoprotein receptor-related protein 5 (LRP5) is essential for normal cholesterol metabolism and glucose-induced insulin secretion. Proc Natl Acad Sci U S A. 2003;100:229–34.
43. Babij P, Zhao W, Small C, et al. High bone mass in mice expressing a mutant LRP5 gene. J Bone Miner Res. 2003;18:960–74.
44. Westendorf JJ, Kahler RA, Schroeder TM. Wnt signaling in osteoblasts and bone diseases. Gene. 2004;341:19–39.
45. Holmen SL, Giambernardi TA, Zylstra CR, et al. Decreased BMD and limb deformities in mice carrying mutations in both Lrp5 and Lrp6. J Bone Miner Res. 2004;19:2033–40.
46. Caira FC, Stock SR, Gleason TG, et al. Human degenerative valve disease is associated with up-regulation of low-density lipoprotein receptor-related protein 5 receptor-mediated bone formation. J Am Coll Cardiol. 2006;47:1707–12.
47. Shao JS, Cheng SL, Pingsterhaus JM, Charlton-Kachigian N, Loewy AP, Towler DA. Msx2 promotes cardiovascular calcification by activating paracrine Wnt signals. J Clin Invest. 2005;115:1210–20.
48. Rajamannan NM. The role of Lrp5/6 in cardiac valve disease: experimental hypercholesterolemia in the ApoE−/− /Lrp5−/− mice. J Cell Biochem. 2011;112:2987–91.
49. Rajamannan NM. Bicuspid aortic valve disease: the role of oxidative stress in Lrp5 bone formation. Cardiovasc Pathol. 2011;20:168–76.

50. Rajamannan NM. The role of Lrp5/6 in cardiac valve disease: LDL-density-pressure theory. J Cell Biochem. 2011;112:2222–9.
51. Rajamannan NM. Oxidative-mechanical stress signals stem cell niche mediated Lrp5 osteogenesis in eNOS(−/−) null mice. J Cell Biochem. 2012;113:1623–34.
52. Rajamannan NM. Embryonic cell origin defines functional role of Lrp5. Atherosclerosis. 2014;236:196–7.
53. Zhao L, Shim JW, Dodge TR, Robling AG, Yokota H. Inactivation of Lrp5 in osteocytes reduces young's modulus and responsiveness to the mechanical loading. Bone. 2013;54:35–43.
54. ten Kate G, Bos S, Dedic A, Neefjes L, Kurata A, Langendonk J. Increased aortic valve calcification in familial hypercholesterolemia: prevalence, extent and associated risk factors in a case-control study. J Am Coll Cardiol. 2015;66:2687–95.
55. Rajamannan NM. Mechanisms of aortic valve calcification: the LDL-density-radius theory: a translation from cell signaling to physiology. Am J Phys Heart Circ Phys. 2010;298:H5–15.
56. Rajamannan NM, Greve AM, Moura LM, Best P, Wachtell K. SALTIRE-RAAVE: targeting calcific aortic valve disease LDL-density-radius theory. Expert Rev Cardiovasc Ther. 2015;13:355–67.
57. Bernoulli D. Hydrodynamica sive de viribus et motibus fluidorum commentarrii. Strasbourg: Argentoratum; 1738. p. St31.
58. Tintut Y, Alfonso Z, Saini T, et al. Multilineage potential of cells from the artery wall. Circulation. 2003;108:2505–10.
59. Jiang M, Wang L, Xuan Q, et al. Risk factors associated with left-sided cardiac valve calcification: a case control study. Cardiology. 2016;134:26–33.
60. Rajamannan NM, Evans FJ, Aikawa E, et al. Calcific aortic valve disease: not simply a degenerative process: a review and agenda for research from the National Heart and Lung and Blood Institute Aortic Stenosis Working Group. Executive summary: calcific aortic valve disease-2011 update. Circulation. 2011;124:1783–91.
61. Wilcox JN, Subramanian RR, Sundell CL, et al. Expression of multiple isoforms of nitric oxide synthase in normal and atherosclerotic vessels. Arterioscler Thromb Vasc Biol. 1997;17:2479–88.

Osteocardiology: The Go/No Go Theory for Clinical Trials

10

Introduction

Recent epidemiological studies have revealed the risk factors associated for vascular atherosclerosis, including male gender, smoking, hypertension and elevated serum cholesterol, are similar to the risk factors associated with development of calcific aortic valve disease (CAVD), calcific aortic disease (CAD) and coronary artery calcification (CAC). The results of the experimental and clinical studies demonstrate that traditional risk factors initiate early atherosclerosis which over time differentiates to form bone in the heart causing, clinical CAVD, CAD, and CAC. It is critical to understand the cellular mechanisms of cardiovascular calcification, the end stage process of the atherosclerosis, to define the critical time point to modify this cellular process before it is too late. Experimental models suggest that medical therapies may have a potential role in patients in the early stages of this disease process to slow the progression of disease. To date, randomized clinical trials in this field have not demonstrated medical therapy can slow progression. Therefore, ongoing studies are necessary to translate cellular biology to turn basic science into future clinical success. This review will summarize the role of Wnt Signaling in osteocardiology to unravel the dilemma of the proper timing of therapy—the Go/No Go time point to slow progression of cardiovascular calcification.

As the global population ages, due to advances in medical therapies, calcific atherosclerotic disease is emerging as a common clinical diagnosis. For years cardiovascular calcification was thought to be due to a degenerative phenomenon by which calcium attaches to the surface of the aortic valve leaflet and the lumen of the vasculature. In 2011, NHLBI recognized that CAVD is an active biologic osteogenic process [1]. Numerous epidemiologic studies were first identified by the Framingham study [2]. The traditional atherosclerotic risk factors include: smoking, male gender, body mass index, hypertension, elevated lipid and inflammatory markers, metabolic syndrome and renal failure [3–16].

N.M. Rajamannan, *Osteocardiology*, DOI 10.1007/978-3-319-64994-8_10

For decades, diagnosing calcification in the heart has been elusive. The advent of computed tomography has opened the window to diagnosing calcification, and calculating the amount of calcification using the Agatston Score [16–18]. Understanding why calcification develops secondary to atherosclerosis in specific locations in the heart which include the coronary artery, left-sided cardiac valves, and the aorta has not been well defined until recently [19]. Understanding the hemodynamic and molecular mechanisms of calcification is critical towards understanding the end-stage calcified phenotype of atherosclerosis or osteocardiology provides the foundation for defining the timing and phenotype expression of bone formation in the heart. The osteocardiology theory correlates experimental evidence with hemodynamic calculations to define the cellular mechanisms of calcification to turn basic science into future clinical success.

Osteocardiology Risk Factors: The Bone-Heart Paradox

For decades, scientific investigations such as the Framingham Heart Study, the Cardiovascular Health Study, and Multi-Ethnic Study of Atherosclerosis, have studied risk factors, which contribute to the pathogenesis of atherosclerosis in the development of cardiovascular disease. Atherosclerosis is a disease in which plaque builds up inside the artery over time. Investigators have determined the risk factors for atherosclerosis utilizing large databases of patients and analyzing the risks associated with specific diagnosis of cardiovascular disease. Over the past 50 years, these large databases have helped to answer several questions important in the understanding of the risk factors and the calcium burden as it relates to, CAVD [20], CAC [21], and CAD [22].

The Multi-Ethnic Study of Atherosclerosis (MESA) study has been instrumental in defining the amount of calcium in the heart, and the associated subclinical risk factors associated with calcification in the heart. Recently, mitral annular calcification (MAC) was also defined in the MESA database [23], as independently associated with cardiovascular risk factors including age, gender, diabetes mellitus, body mass index, status of current smoking and use of lipid lowering therapy. These important databases, in addition to defining the calcium burden in the heart as measured by CT scan, have identified novel risk factors such as Lp(a), which is specific to causing CAVD [20, 24]. All of these studies, have demonstrated that traditional atherosclerotic risk factors are in part responsible for the development of CAVD, CAD, CAC, and MAC, associated with variable calcification expression depending on the anatomic location. Defining the osteocardiology phenotype recognizes that in the presence of these traditional risk factors, calcification can develop in specific locations in the heart which include: the left-sided heart valves, the aorta, and the coronary artery.

In addition, large epidemiologic databases have further developed the concept that atherosclerosis and osteoporosis develops simultaneously secondary to traditional cardiovascular risk factors [25]. Calcification in the heart and osteoporosis in the bone is a common diagnosis in the aging population. The paradox of bone formation in the heart and thinning bone in the femur secondary to atherosclerosis has been confirmed in a LDLR null mouse model [26]. Understanding of the parallel role of bone in the heart is becoming increasing important since the phenotype of calcification in the valve is similar to an osteogenic process [27]. This paradox provides a foundation for the theory correlating risk factors, epidemiology, disease mechanisms and possibility for medical therapy.

Translating Experimental Studies into Understanding Osteocardiology: The LDL-Density-Radius Theory

Lessons from the experimental studies have evolved into a series of clinical parameters, which provide the foundation for an algorithm to treat CAVD, CAC and CAD: the LDL-Density-Radius Theory [28]. From a valve and vascular biologist perspective, the possibility for medical therapy for osteocardiology resides in two fundamental differences in vascular versus valvular biology: first is calculating the magnitude of LDL lowering necessary to treat the process, and second is the difference in the radius between the aortic valve and that of the vessel. These differences are important to understand for the final analysis of these trials, and for the future trial design for osteocardiology.

The hypothesis to measure the effects of lipid lowering in slowing the progression of osteocardiology is dependent on two axioms, biology and hemodynamics. The experimental data demonstrates that lipids activate the bone differentiation within atheroma in the valve and in the vessel. The first axiom is the LDL density theory. The first axiom accounts for the effect of low-density lipoprotein (LDL) biology in atherosclerosis. If the risk factors of elevated cholesterol and LDL are important in this disease then measuring lipid lowering using standard established assays for LDL in the treatment of valve disease becomes necessary. This approach does not take into account the effect of other inflammatory contributors to this disease including Lp(a) [20, 24], and other inflammatory markers, which also contribute to the pathogenesis of valvular and vascular disease but are not routinely measured in everyday clinical practice.

The direction of the LDL affects the vascular lumen in an inward direction causing occlusion overtime, as shown in Fig. 10.1, Panel a. The direction of this LDL affects the valve is an upward direction along the y-axis along the aortic surface of the valvular fibrosa. Overtime, the leaflets stiffen and can fuse in some valves. The overall effect on the radius is a reduction in the aortic valve opening and obstruction which leads to progressive stenosis of the valve and calcification of the aorta as shown in Fig. 10.1, Panel b. Figure 10.1, Panel c, demonstrates a formula to calculate the percent reduction of the LDL density before and after therapy similar to the calculation derived in the Reversal trial measuring reductions in atheroma volume in coronary artery disease [29]. Calculation of the percent lowering of LDL density in a valve trial allows for the potential to calculate the improvement on the biologic effect of LDL on this disease.

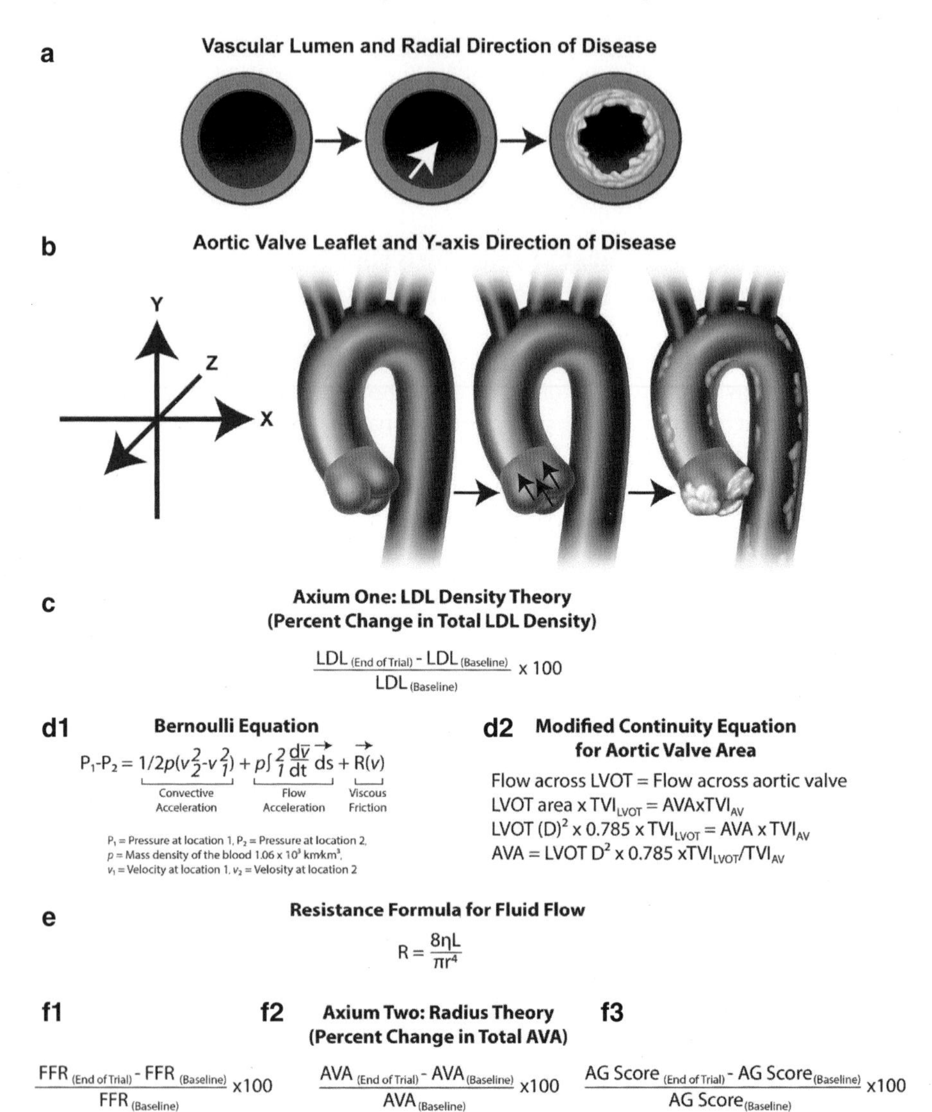

Fig. 10.1 The LDL-Density-Radius Theory [44] (Permission obtained for reproduction). Panel **a** vascular lumen and radial direction of disease. Panel **b** aortic valve leaflet and Y-axis direction of disease. Panel **c** axiom one: LDL-density theory. Panel **d1** Bernoulli equation. Panel **d2** modified continuity equation for aortic valve area. Panel **e** Resistance for Fluid Flow. Panel **f** axiom two: radius theory, **f1** FFR in CAC; **f2** AVA in CAVD; **f3** AG score in CAD

The second axiom for the theory is the radius theory. This hemodynamic radius principle is based on the biologic direction of this disease. The second axiom calculates the biologic effect of the changes in the radius for specific the anatomic location in the heart. Fig. 10.1, Panel d1, the formula for Bernoulli flow through a pipe, as modified [30] for echocardiography. Figure 10.1, Panel d2, is the formula to calculate aortic valve areas by echocardiography using the doppler technique [31]. The derivation of the Bernoulli Principal for this equation includes the drop of the calculation for the flow acceleration and the viscous friction because the velocity profile in the center of the lumen is usually so low that the effect of viscous friction becomes insignificant and not necessary to calculate. Clinically, the viscous friction factor has been ignored as part of the continuity equation in aortic valve disease as defined by the echocardiography physiologists [32].

However, the concept of viscous friction becomes important when comparing vascular trials to valvular trials. The size of the radius plays a very important role in the time to see treatment effects, which are defined by vascular clinical end-points such as ischemia and acute myocardial infarction. Clinical results from the trial entitled FAME [33], revealed the most stringent results for the role of Fractional Flow Reserve (FFR) in the diagnosis of physiologic critical stenosis in coronary artery disease in lesions. FFR as the continuity equation measure flow via pressure differential versus velocity differential as derived from the Bernoulli Equation. In 20 centers in Europe and the United States, 1005 patients undergoing percutaneous coronary intervention with stent implantation, were randomized based on angiography or based on FFR in addition to angiography. Results demonstrated that improved endpoints in the FFR group with less stent use in patients with angiographic significant lesions and a FFR less than 0.80.

To date SEAS and SALTIRE, randomized clinical trials in valvular heart disease, were designed using the vascular trialists approach, which resulted in negative results in slowing progression of CAVD [34]. However, because the flow in the lumen of the vasculature is not flat due to a smaller radius [35], the viscous friction factor must be taken into account in evaluating the treatment effects within the vasculature as derived by Bernoulli's original equation [35]. Therefore, the treatment effect of LDL lowering will have a more rapid effect on the vasculature as compared to the heart valve.

The importance of the smaller radius is shown in Fig. 10.1, Panel e, which is the calculation of resistance of fluid through a pipe. If the size of the radius(r) is significant in the calculation of flow, then the inverse r^4 dependence of the resistance becomes important in the treatment a smaller radius versus a larger radius in the aortic valve area as viscosity increases by a factor of 16. Therefore, comparing the rates of improvement in a vascular trial versus a valvular trial will be different due the differences in the size of the radius and the derivation of the modified Bernoulli equation for the echocardiographic formula for valve areas. The Continuity Equation drops the calculation of viscous friction due to the large size of the radius of the outflow tract of the left ventricle.

To measure the treatment effect for coronary artery disease: Fig. 10.1, Panel f1, is the calculation for the percent improvement for the FFR. To measure the treatment effect for aortic valve disease: Fig. 10.1, Panel f2, is the calculation for percent improvement for AVA. To measure the treatment effect for aortic disease: Fig. 10.1, Panel f3, is the calculation for the percent improvement for aortic disease. Mathematically and biologically, clinical trials for aortic valve disease may consider the following two axioms for targeting the disease biology in terms of the radial direction of disease and the magnitude of the LDL density to activate the atherosclerotic process according to Bernoulli's original formula and the effect on resistance and fluid flow. The effect will require a longer period of time to see slowing of progression in the Aortic valve area for the reasons described in the LDL-Density-Radius Theory. Furthermore, the effect may be masked in the results of the published trials as the patients were randomized to treatment and the 2 axioms described in this theory are not accounted for in the randomization protocol.

Timing of Therapy: Osteocardiology Go/No Go Theory

The timing of treatment to slow progression of calcification in the heart has been difficult to achieve with randomized clinical trials. Clinical trials targeting calcification in the coronary artery, aortic valve and aorta have been variable, but for the most part negative in part for several reasons. First for years this disease was thought to be due to a degenerative process secondary to passive calcification in the heart. A full understanding of the biology of atherosclerotic calcification in the heart will help to understand the initiation, early atherosclerosis manifesting in subclinical disease, and late calcification manifesting in severe clinical disease. The concept of identifying and treating early pre-clinical atherosclerosis-versus late calcification defines the principle of go/no go binary classification of the disease compendium. Figure 10.2, demonstrates from a biologic and clinical perspective the Go/No Go timing for the treatment of calcification in the cardiovascular system. Treat the modifiable disease while it is treatable in a "*Go*" state, versus working on treatments before it is too late—severe calcification or the "*No Go*" state. The trial will be positive only when the *Go* condition is met, and also when the *No Go* condition fails such as the randomized control trials in aortic valve disease [36–38]. In the future, the design of clinical trials in atherosclerotic heart disease, needs to focus on the early subclinical phase of atherosclerosis—the Go phase of disease, to try and reverse atherosclerosis as found in the Reversal Trial.[51] Once calcification starts then medical therapies may have the possibility of halting progression, but the clinical trials in calcific aortic valve disease, demonstrate that the disease progressed despite intensive lipid lowering therapy [36–38].

MESA Predictions

MESA has become invaluable in describing subclinical risk factors and the association of calcification in the heart. In summary, there have been over 1200 publications from MESA research, however, the critical discoveries in the role of calcification and the timing, risk factor in subclinical atherosclerosis will provide the foundation for the timing of the Go/No Go theory of clinical trials in the field of osteocardiology.

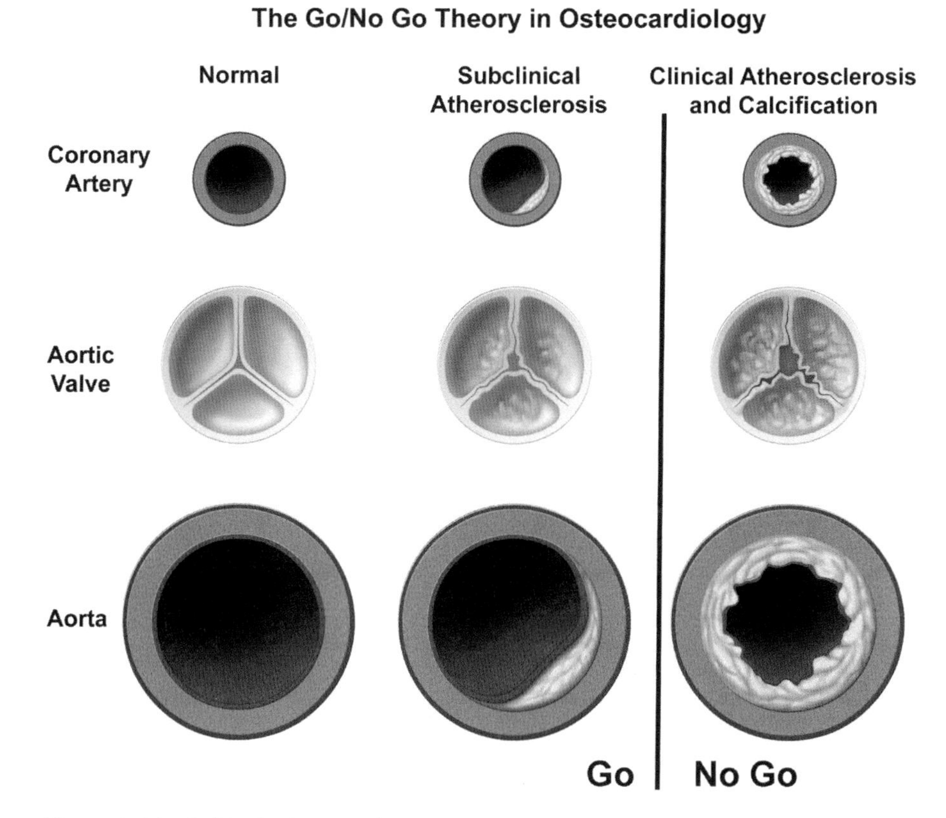

Fig. 10.2 The Go/No Go Osteocardiology Theory: Timing to treat calcification in the cardiovascular system while the disease process is in the subclinical stage of atherosclerosis prior to the development of severe disease and calcification [44] (Permission obtained for reproduction)

CAC

1) The prevalence and 75th percentile levels of CAC were highest in white males and lowest in African American and Hispanic females.
2) In addition, a recent cost-effectiveness analysis based on data from MESA reported that CAC testing and statin treatment for those with CAC > 0 was cost effective in intermediate-risk scenarios.
3) Furthermore, a recent MESA analysis compared these CAC-based treatment strategies to a "treat all" strategy and to treatment according to the ATPIII guidelines with clinical and economic modeled over both 5- and 10-year time horizons. The results consistently demonstrated that it is both cost-saving and more effective to scan intermediate-risk patients for CAC and to treat those with CAC ≥ 1 that to use treatment based on established risk assessment guidelines.
4) Atherosclerosis is a chronic, progressive, inflammatory disease with a long asymptomatic phase. The long asymptomatic phase of the disease process, is a

critical time point for identifying risk factors, initial stages of disease and any sign of early calcification to treat, modify and try to halt, slow or reverse progression. race/ethnicity [39].
5) The presence of existing coronary artery calcification did not affect these associations of Lp(a) and CAVD.

CAVD

1) The risk factors in MESA associated with the newly diagnosed CAVD, included age, male gender, body mass index, current smoking, and the use of lipid-lowering and antihypertensive medications.
2) Among those with CAVD at baseline, the median rate calcification progression was 2 Agatston units/year [40]. The baseline Agatston score was a strong, independent predictor of progression, especially among those with high calcium scores at baseline.
3) In conclusion, in this MESA, preclinical cohort, the rate of incident CAVD increased significantly with age.
4) MESA also confirmed the discovery of Lp(a) as a significant risk factor for CAVD [20].
5) Importantly, they found that aortic valve calcium independently predicts coronary and cardiovascular events in a primary prevention MESA population [41].

CAD

1) MESA is the first to determine parallel risk factors for traditional cardiovascular risk factors and risk factors for CAD, with a long term subclinical phase until it reaches clinical overt disease.
2) The study indicates that CAD is a significant predictor of future coronary events only in women, independent of coronary artery calcification (CAC) [42].
3) MESA confirmed the independent association between volumetric trabecular bone mineral density (vBMD) of the lumbar spine and coronary artery calcification (CAC) and calcific aortic disease (CAD) [43].
4) The Montreal Group also defined the extent of calcific aortic disease in the FH population, including the first to describe a gene dosage effect [22] and an association with age for progression.

Summary

For the past 50 years, catheter hemodynamics, echocardiography, angiography, CT imaging and timing of surgery have evolved as the diagnostic and therapeutic approach for CAVD, CAC, MAC and CAD. In the past decade, with the advent of experimental models and genetic testing, recognition that the aortic valve has an active cellular biology which incorporates risk factors, osteogenic phenotype and the potential for medical therapy to slow progression. The Wnt signaling pathway activates mesenchymal cell differentiation in the valve and vasculature for the development of an osteoblast phenotype. The future management of this disease process will include the understanding of these different mechanisms for future medical therapy of this disease. If the physician can define the traditional risk factors in patients who present with an aortic valve murmur then targeting these risk factors may slow progression. The stethoscope can become an inexpensive screening tool for this pathologic process and possible subclinical atherosclerosis. If there are no identifiable risk factors then genetic considerations may play a role. Progress in this field, will make a difference for the future delay in the timing of surgery for these patients in the future. In 2017, incorporation of risk factor evaluation, diagnosis of subclinical and clinical disease, and the overall health of the patient in future cardiovascular risk management will be the most important clinical and scientific approach towards treatment of our patients. Understanding the biology of calcification will help to understand the timing, the future clinical trial design and duration of therapy to achieve success in treating osteocardiology secondary to atherosclerosis.

References

1. Rajamannan NM, Evans FJ, Aikawa E, et al. Calcific aortic valve disease: not simply a degenerative process: a review and agenda for research from the National Heart and Lung and Blood Institute Aortic Stenosis Working Group. Executive summary: calcific aortic valve disease-2011 update. Circulation. 2011;124:1783–91.
2. FHS. Framingham Heart Study Website. 2017. http://wwwframinghamheartstudyorg. Accessed 2 March 2017.
3. Deutscher S, Rockette HE, Krishnaswami V. Diabetes and hypercholesterolemia among patients with calcific aortic stenosis. J Chronic Dis. 1984;37:407–15.
4. Hoagland PM, Cook EF, Flatley M, Walker C, Goldman L. Case-control analysis of risk factors for presence of aortic stenosis in adults (age 50 years or older). Am J Cardiol. 1985;55:744–7.
5. Aronow WS, Ahn C, Kronzon I, Goldman ME. Association of coronary risk factors and use of statins with progression of mild valvular aortic stenosis in older persons. Am J Cardiol. 2001;88:693–5.
6. Mohler ER, Sheridan MJ, Nichols R, Harvey WP, Waller BF. Development and progression of aortic valve stenosis: atherosclerosis risk factors—a causal relationship? A clinical morphologic study. Clin Cardiol. 1991;14:995–9.
7. Lindroos M, Kupari M, Valvanne J, Strandberg T, Heikkila J, Tilvis R. Factors associated with calcific aortic valve degeneration in the elderly. Eur Heart J. 1994;15:865–70.
8. Boon A, Cheriex E, Lodder J, Kessels F. Cardiac valve calcification: characteristics of patients with calcification of the mitral annulus or aortic valve. Heart. 1997;78:472–4.
9. Chui MC, Newby DE, Panarelli M, Bloomfield P, Boon NA. Association between calcific aortic stenosis and hypercholesterolemia: is there a need for a randomized controlled trial of cholesterol-lowering therapy? Clin Cardiol. 2001;24:52–5.
10. Wilmshurst PT, Stevenson RN, Griffiths H, Lord JR. A case-control investigation of the relation between hyperlipidaemia and calcific aortic valve stenosis. Heart. 1997;78:475–9.
11. Chan KL, Ghani M, Woodend K, Burwash IG. Case-controlled study to assess risk factors for aortic stenosis in congenitally bicuspid aortic valve. Am J Cardiol. 2001;88:690–3.
12. Briand M, Lemieux I, Dumesnil JG, et al. Metabolic syndrome negatively influences disease progression and prognosis in aortic stenosis. J Am Coll Cardiol. 2006;47:2229–36.

13. Palta S, Pai AM, Gill KS, Pai RG. New insights into the progression of aortic stenosis: implications for secondary prevention. Circulation. 2000;101:2497–502.
14. Peltier M, Trojette F, Sarano ME, Grigioni F, Slama MA, Tribouilloy CM. Relation between cardiovascular risk factors and nonrheumatic severe calcific aortic stenosis among patients with a three-cuspid aortic valve. Am J Cardiol. 2003;91:97–9.
15. Stewart BF, Siscovick D, Lind BK, et al. Clinical factors associated with calcific aortic valve disease. Cardiovascular Health Study. J Am Coll Cardiol. 1997;29:630–4.
16. Pohle K, Maffert R, Ropers D, et al. Progression of aortic valve calcification: association with coronary atherosclerosis and cardiovascular risk factors. Circulation. 2001;104:1927–32.
17. Messika-Zeitoun D, Aubry MC, Detaint D, et al. Evaluation and clinical implications of aortic valve calcification measured by electron-beam computed tomography. Circulation. 2004;110:356–62.
18. Bild DE, Detrano R, Peterson D, et al. Ethnic differences in coronary calcification: the multiethnic study of atherosclerosis (MESA). Circulation. 2005;111:1313–20.
19. Lazaros G, Toutouzas K, Drakopoulou M, Boudoulas H, Stefanadis C, Rajamannan N. Aortic sclerosis and mitral annulus calcification: a window to vascular atherosclerosis? Expert Rev Cardiovasc Ther. 2013;11:863–77.
20. Cao J, Steffen BT, Budoff M, et al. Lipoprotein(a) levels are associated with subclinical calcific aortic valve disease in White and Black individuals: the multi-ethnic study of atherosclerosis. Arterioscler Thromb Vasc Biol. 2016;36:1003–9.
21. Ten Kate GJ, Neefjes LA, Dedic A, et al. The effect of LDLR-negative genotype on CT coronary atherosclerosis in asymptomatic statin treated patients with heterozygous familial hypercholesterolemia. Atherosclerosis. 2013;227:334–41.
22. Awan Z, Alrasadi K, Francis GA, et al. Vascular calcifications in homozygote familial hypercholesterolemia. Arterioscler Thromb Vasc Biol. 2008;28:777–85.
23. Hamirani YS, Nasir K, Blumenthal RS, et al. Relation of mitral annular calcium and coronary calcium (from the Multi-Ethnic Study of Atherosclerosis [MESA]). Am J Cardiol. 2011;107:1291–4.
24. Thanassoulis G, Campbell CY, Owens DS, et al. Genetic associations with valvular calcification and aortic stenosis. N Engl J Med. 2013;368:503–12.
25. Figueiredo CP, Rajamannan NM, Lopes JB, et al. Serum phosphate and hip bone mineral density as additional factors for high vascular calcification scores in a community-dwelling: the Sao Paulo Ageing & Health Study (SPAH). Bone. 2013;52:354–9.
26. Rajamannan NM. Atorvastatin attenuates bone loss and aortic valve atheroma in LDLR mice. Cardiology. 2015;132:11–5.
27. Rajamannan NM, Edwards WD, Spelsberg TC. Hypercholesterolemic aortic-valve disease. N Engl J Med. 2003;349:717–8.
28. Rajamannan NM. Mechanisms of aortic valve calcification: the LDL-density-radius theory A: translation from cell signaling to physiology. Am J Physiol. 2010;298:H5–15.
29. Nissen SE, Tuzcu EM, Schoenhagen P, et al. Effect of intensive compared with moderate lipid-lowering therapy on progression of coronary atherosclerosis: a randomized controlled trial. JAMA. 2004;291:1071–80.
30. Hatle L, Angelsen BA, Tromsdal A. Non-invasive assessment of aortic stenosis by Doppler ultrasound. Br Heart J. 1980;43:284–92.
31. Smith MD, Kwan OL, DeMaria AN. Value and limitations of continuous-wave Doppler echocardiography in estimating severity of valvular stenosis. JAMA. 1986;255:3145–51.
32. Hatle L, Brubakk A, Tromsdal A, Angelsen B. Noninvasive assessment of pressure drop in mitral stenosis by Doppler ultrasound. Br Heart J. 1978;40:131–40.
33. Tonino PA, De Bruyne B, Pijls NH, et al. Fractional flow reserve versus angiography for guiding percutaneous coronary intervention. N Engl J Med. 2009;360:213–24.
34. Rajamannan NM. Calcific aortic valve disease in familial hypercholesterolemia: the LDL-density-gene effect. J Am Coll Cardiol. 2015;66:2696–8.
35. Bernoulli D. Hydrodynamica sive de viribus et motibus fluidorum commentarrii. Strasbourg: Argentoratum; 1738. p. St31.

36. Gerdts E, Rossebo AB, Pedersen TR, et al. Impact of baseline severity of aortic valve stenosis on effect of intensive lipid lowering therapy (from the SEAS study). Am J Cardiol. 2010;106:1634–9.
37. Cowell SJ, Newby DE, Burton J, et al. Aortic valve calcification on computed tomography predicts the severity of aortic stenosis. Clin Radiol. 2003;58:712–6.
38. Chan KL, Teo K, Dumesnil JG, Ni A, Tam J, Investigators A. Effect of Lipid lowering with rosuvastatin on progression of aortic stenosis: results of the aortic stenosis progression observation: measuring effects of rosuvastatin (ASTRONOMER) trial. Circulation. 2010;121:306–14.
39. McClelland RL, Jorgensen NW, Budoff M, et al. 10-year coronary heart disease risk prediction using coronary artery calcium and traditional risk factors: derivation in the MESA (Multi-Ethnic Study of Atherosclerosis) with validation in the HNR (Heinz Nixdorf Recall) study and the DHS (Dallas Heart Study). J Am Coll Cardiol. 2015;66:1643–53.
40. Owens DS, Katz R, Takasu J, Kronmal R, Budoff MJ, O'Brien KD. Incidence and progression of aortic valve calcium in the multi-ethnic study of atherosclerosis (MESA). Am J Cardiol. 2010;105:701–8.
41. Owens DS, Budoff MJ, Katz R, et al. Aortic valve calcium independently predicts coronary and cardiovascular events in a primary prevention population. JACC Cardiovasc Imaging. 2012;5:619–25.
42. Budoff MJ, Nasir K, Katz R, et al. Thoracic aortic calcification and coronary heart disease events: the multi-ethnic study of atherosclerosis (MESA). Atherosclerosis. 2011;215:196–202.
43. Hyder JA, Allison MA, Wong N, et al. Association of coronary artery and aortic calcium with lumbar bone density: the MESA Abdominal Aortic Calcium Study. Am J Epidemiol. 2009;169:186–94.
44. Rajamannan N. Wnt signaling in atherosclerosis: the Go/No Go theory for clinical trials in osteocardiology. Cardiology. 2017. (in press).

Printed by Printforce, the Netherlands